CONSTRUCTION MATERIALS AND ENGINEERING

OIL PALM SHELL

LIGHTWEIGHT CONCRETE AND STRUCTURES

Construction Materials and Engineering

Additional books in this series can be found on Nova's website under the Series tab.

Additional e-books in this series can be found on Nova's website under the eBooks tab.

CONSTRUCTION MATERIALS AND ENGINEERING

OIL PALM SHELL

LIGHTWEIGHT CONCRETE AND STRUCTURES

MOHAMMAD ABDUL MANNAN

nova
science publishers
New York

Library of Congress Cataloging-in-Publication Data

ISBN: 978-1-53610-878-1

Published by Nova Science Publishers, Inc. † New York

Al Qur'an: Sura Hud (11/46), "He said: O Nūh!........
I give you counsel, lest you act like the ignorant!"

Dedication of this book to:

My beloved wife, Firoza Khatun
My lovely daughter, Sadia Tasnim and
My lovely son, Mohammad Ridwan

CONTENTS

PREFACE

Due to a high demand in construction and furniture industries worldwide, natural resources such as stones and wood as non-renewable resources are being depleted. Thus, researchers are focusing on renewable resources as alternative materials. As such, the utilization of abundant solid wastes and byproducts, which are discharged from agriculture, industry and municipalities present an alternative to the conventional materials for the construction and furniture industries. These solid wastes and byproducts, when properly processed have shown to be effective and can readily meet design specifications. Agricultural solid wastes from oil palm distributors such as Oil Palm Shell (OPS) and Empty Fruit Bunch (EFB), which are abundant in agro-based countries, present an interesting alternative to the conventional aggregate in lightweight concrete and artificial plank production, respectively.

At present, palm oil producing countries are Barkina Faso, Benin, Burundi, Cameroon, Central African Republic, Colombia, Costa Rica, Côte d'Ivoire, Democratic Republic of Congo, Ecuador, Equatorial Guinea, Gabon, Gambia, Ghana, Guinea Bissau, Guinea, Honduras, India, Indonesia, Liberia, Malaysia, Mexico, Nigeria, Papua New Guinea, Peru, Republic of Congo, Senegal, Sierra Leone, Tanzania, Thailand, Togo, Uganda, Venezuela and others.

In Malaysia, oil palm plantations cover over 5 million hectares, and annual production of OPS as solid waste from 450 oil palm mills is more than 6 million tons. This large amount of OPS as a renewable green aggregate can contribute to overcoming the over dependence on depletable resources for concrete production. The civil engineering projects are of a larger scale; they need sustainable materials in order to gain a greater momentum of growth. The major technical characteristics of OPS solid waste must be primarily understood before each particular use. Therefore, there is a need to highlight the importance of OPS to be used in the construction industry.

Hence, the book is covered with 10 Chapters. A brief explanation of each chapter is given. Chapter 1 covers the generation of solid wastes in the Oil palm industry and the selection criteria of solid wastes. This chapter also demonstrates various potential applications of concrete and artificial plank using these solid wastes. Chapter 2 covers physical, mechanical and chemical properties of OPS and different pre-treatments on OPS. Chapter 3 highlights the resistance capacity of OPS aggregate in acid, salt and alkali environments. Chapter 4 elaborates the mix procedure of OPS concrete as lightweight concrete. Chapter 5 explains fresh, hardened and durability properties of OPS concrete. Chapter 6 explains flexural behaviour of reinforced OPS concrete beam. Chapter 7 describes the flexural behaviour of reinforced OPS concrete one-way slabs, especially precast C-channel as the most efficient product for building floors. Chapter 8 explains the long-term performance of prototype structures of 14, 12 and 6 years old built with OPS concrete. Chapter 9 explains how, in precast concrete components in the industrialised building system, IBS plays multifunctionally on decorative retaining walls, water seepage drains and slope surface protection. Mosquitoes cause serious problems to humans and other animals globally. They breed in standing shallow water particularly in summer monsoon rain that stands more than three days. As such, open drains in flat land cause water stagnant favouring mosquito breeding. The best way to eliminate the breeding sites of mosquitoes can be done using water seepage drains as explained in this chapter. Lastly, Chapter 10 explains o how houses and eco-resorts in marine environments can be built using precast concrete components.

DISTINGUISHING FEATURES OF THE BOOK

The approaches mentioned above are new and will be useful for researchers, the oil palm and construction industries, practitioners etc. The oil palm industry will generate additional revenue through value added products, namely high density planks, lightweight concrete etc. using the downstream solid wastes as discharged from oil palm mills. The particle board industry may also look into high-density plank production. The researchers through this book will enhance knowledge for the investigation. The precast SIConSofa as a water seepage drain and retaining wall system is useful for municipality and housing development sectors. This book brings business opportunity for the precast industry on new types of cost efficient precast products. The marine ecotourism sector may look into using full precast structural components for resort construction.

ABOUT THE AUTHOR

Dr. Mohammad Abdul Mannan was born to a simple family of a small village, Aktarpur, Rangiarpota, Jibonnagar, Chuadanga, Bangladesh. He obtained B.Sc. (Civil Engineering) with first class, MSc in Civil Engineering and a PhD in Concrete technology. He has started his career as a lecturer at BIT Rajshahi (now RUET), Bangladesh followed by AJP consulting firm, then Universiti Malaysia Sabah (UMS) and is now a Professor in the Department of Civil Engineering, Universiti Malaysia Sarawak, Malaysia. He is the inventor of a few construction products. Based on 30 years of experience in teaching, professional practice and research, his vision is to be excellent in research on Innovative Construction Material and Structure. Dr. Mannan can be reached at mannan27@hotmail.com and mannan@unimas.my

ACKNOWLEDGMENTS

The author through this book remembers his family, relatives, teachers, colleagues, students and friends sincerely. Particularly, the author acknowledges his old student Ashley Anak Eng Bee for helping with some figures in this book. The author also expresses thanks to Dr. C. H. Ng and Mrs. Hidayati Asrah for providing necessary information useful in this book. .

Chapter 1

SOLID WASTES IN THE OIL PALM INDUSTRY

Keywords: Oil palm industry, Solid wastes as renewable resource, Oil palm shell (OPS), Empty fruit bunch (EFB), Nut fibers, Oil palm boiler clinker (OPBC), Palm oil fuel ash (POFA), Solid waste evaluation criteria

1.1. INTRODUCTION

The natural stones of both high and marginal quality as non-renewables are being depleted due to high construction demand worldwide. Thus, the researchers are focusing on renewable resources as alternative construction materials. As such, the use of abundant solid wastes and byproducts, which are discharged from agriculture, industry and municipalities present an alternative to conventional construction materials. These solid wastes and byproducts, when properly processed have shown to be effective and can meet the specifications.

At present, palm oil producing countries are Barkina Faso, Benin, Burundi, Cameroon, Central African Republic, Colombia, Costa Rica, Côte d'Ivoire, Democratic Republic of Congo, Ecuador, Equatorial Guinea, Gabon, Gambia, Ghana, Guinea Bissau, Guinea, Honduras, India, Indonesia, Liberia, Malaysia, Mexico, Nigeria, Papua New Guinea, Peru, Republic of Congo, Senegal, Sierra Leone, Tanzania, Thailand, Togo, Uganda, Venezuela and others.

In Malaysia, the annual discharge of solid waste from 450 oil palm mills [1] is in large quantity. The solid wastes discharged daily such as oil palm shell (OPS), nut fibers, empty fruit bunch (EFB), oil palm boiler clinker (OPBC) and palm oil fuel ash (POFA) at palm oil mills are considered renewable materials (Table 1.1) particularly for the construction industry, furniture industry etc. It is estimated that the oil palm mills having crushing capacity of 45 tons Fresh Fruit Bunch (FFB) hourly is able to discharge 43.2 tons of OPS and 29.52 tons of dry EFB fibers as shown in Tables 1.2 and 1.3 daily. About 6.29 tons OPS is required to construct one low-cost house having built up area of 58.68m^2. Similarly, about 25kg dry EFB fibers are required to make an artificial plank of 1200mm×2400mm×12mm in size. Thus, these huge amounts of OPS and EFB as renewable material can contribute to overcoming the over-dependence on depletable natural resources such as stone and wood for concrete and plank production respectively.

The civil engineering projects need sustainable materials in order to gain greater momentum of growth. The major technical characteristics of OPS solid waste must be primarily understood before each particular use. Thus, the physical, mechanical and chemical properties of such solid wastes must be investigated in order to satisfy the technical specification, economic gain, social and environmental benefits as sustainable products for total gain in a particular application. It is also essential to investigate the concrete made using such solid waste, whether lightweight brings advantages in relation to gravitation and seismic forces for building. The social benefit can be observed through overall economic benefits including conservation of natural resources and environmental purification. It is also essential to measure the adverse impacts such as possible losses and environmental degradation of using such solid waste.

1.2. Generation of Solid and Liquid Wastes in the Oil Palm Industry

Oil palm is an economically important crop in several humid and tropical countries. Malaysia is one such zone of warm humid climate [2]. At oil palm plantations, solid wastes are in the form of leaves, fronds and trunks. Trunks are obtained once trees are cut down. At oil palm mills, the solid wastes discharged are as:

a. Empty fruit bunch (EFB)
b. Oil palm shell (OPS
c. Nut fibers (pericarp)
d. Oil palm boiler clinker (OPBC) and
e. Palm oil fuel ash (POFA)

The effluent is also discharged at the mill. Both solid and liquid wastes are discharged daily from oil palm mills as renewable resources and their potential usage is shown in Table 1.1. These solid wastes bring economic and environmental benefits, and oil palm mills get additional income. Figure 1.1 shows the heap of fresh fruit bunches (FFB). The use of such agricultural solid wastes in the construction industry, furniture industry etc., particularly concrete making has engineering potential and economic advantages. The estimated methane gas emission from the oil palm industry in Malaysia is about 0.21 million tons [3]. This methane gas is mostly from unmannered disposal of excess EFB, OPS and nut fibres at mill sites and is seen self-burning (Figure 1.2).

Table 1.1. Renewable wastes discharged from Oil palm mill and potential usage

No.	Name of waste	Potential usage for
1	Oil palm shell (OPS)	• Aggregate for lightweight concrete • High density plank • Fuel
2	Oil palm boiler clinker (OPBC)	• Aggregate for lightweight concrete
3	Palm oil fuel ash (POFA)	• Supplementary cementitious material (SCM) for partial cement replacement in concrete

No.	Name of waste	Potential usage for
4	Empty fruit bunch (EFB) fibre	• Animal feed supplement • Boiler fuel • Erosion control mat • Fertilizer mat • Fibre reinforced concrete • Food container • High density plank • Horticulture • Mattress • Mulching for landscaping, young plants and seedlings • Power generation • Pulp and Paper • Soil stabilization • Others
5	Nut fibre	
6	Palm oil mill effluent (POME)	• Methane extraction for fuel

Figure 1.1. Heap of Fresh Fruit Bunches (FFB) at an Oil palm mill (courtesy, author).

Figure 1.2. Self-burning and unmannered disposal of solid wastes such as EFB, nut fibers etc. at an Oil palm mill (courtesy, Lau Pei Ching, personal contact).

1.2.1. Oil Palm Shell

Generally, an oil palm mill having an hourly FFB crushing capacity of 45 tons is able to discharge about 43.2 tons OPS daily (Table 1.2). It is used partly for fuel in boilers and the excess is disposed of (Figure 1.3). The actual shape and size of OPS is shown in Figure 1.4 which is considered similar to the conventional crushed aggregate used for concrete. A single story low-cost house of 58.68m^2 built up area needs about 17m^3 concrete in which 6.29 tons OPS will act as a coarse aggregate. Use of OPS as an aggregate in lightweight concrete brings advantages in terms of the self-weight reduction of structural components using such concrete against gravitational and seismic forces. The use of OPS as a coarse aggregate in concrete making is the main purpose of this book.

Table 1.2. Estimation of OPS discharge from one oil palm mill and use of OPS in concrete for low-cost house construction (courtesy, author)

Description	Quantity
Assumed hourly crushing capacity with Fresh Fruit Bunch (FFB) of one palm oil mill	45 tons/hour
Daily operation time of crusher	16 hours/day
Daily total crushing capacity with FFB (45tons/hour×16 hours=720 tons)	720 tons/day
Oil Palm Shell (OPS) aggregate production from 100 tons FFB	**6 tons OPS**
Daily total OPS disposed [(6 tons OPS/100 tons FFB) ×720 tons FFB=43.2 tons]	**43.2 tons OPS/day**
Annual OPS disposed by one mill=43.2tons×365days=15,768 tons×70% efficiency (assumed)=11,037 tons/year/mill	**11,037 tons OPS/year**
Required amount of OPS aggregate for OPS concrete, 28-day comp. st of 24.20N/mm^2 [4]	370kg OPS for 1 m^3 concrete
Required OPS for one unit low-cost house construction=17m^3 concrete/unit×370kg OPS/m^3 concrete=6290kg=6.29tons	6.29tons OPS/unit low-cost house
Daily concrete production using 43.2 tons OPS aggregate= 43.2×1000kg÷370kg=116.76m^3	116.76m^3 concrete /day
Annual concrete production =116.76×365 days=42,617m^3 units×70% optimization (assumed)=29,832m^3	29,832m^3 concrete/year/mill
No. of low-cost houses to be built using 29832m^3 OPS concrete =29832m^3÷17m^3/unit (assumed) =1754 =1,700 (rounded) units [size of low-cost house =58.68m^2 built up area for 2 beds, 1 study, 1 living cum dining, 1 kitchen, 1 shower, 1 toilet and 1 laundry space]	**1,700 units low-cost house/year/Oil palm mill**

Figure 1.3. Large heap of disposed OPS at an oil palm mill [courtesy, Lau Pei Ching, personal contact].

Figure 1.4. Resulting shapes and sizes of OPS through breaking of nuts for extraction of karnels [courtesy, author].

1.2.2. Nut Fibers

Figure 1.5 shows a heap of nut fibers. They are very short and narrow, and relatively weak in tension compared with other fibers. It is primarily used as fuel in boilers. These nut fibers can be used as value added product for other applications.

Figure 1.5. Heap of disposed nut fibers at an oil palm mill [courtesy, Faris Nur, personal contact].

1.2.3. Palm Oil Mill Effluent and Sludge

Figure 1.6 shows the POME (palm oil mill effluent) ponds where palm oil mill effluent is discharged. This pond is the source of Methane gas and this gas can be obtained from this pond. The sludge must be investigated for possible use as supplementary cementitious material (SCM) for partial cement replacement in concrete making.

Figure 1.6. Effluent (POME) ponds near an oil palm mill [courtesy, author].

1.2.4. Oil Palm Boiler Clinker and Palm Oil Fuel Ash

The nut fibers and OPS are used as fuel for boilers at oil palm mills. Sometimes, small EFBs are also used for fuel. The resultant residues are known as palm oil fuel ash (POFA) and hard lump known as oil palm boiler clinker (OPBC) as shown in Figures 1.7 to 1.10. In most cases, POFA is discharged in a black color due to incomplete burning. Usually, inconsistent feeding to a boiler is done manually using nut fibers, OPS and sometimes EFB, and the resultant residue is black color POFA. The feeding can enhance proper burning and the residue (OPBC and POFA) appears as off white as shown in Figure 1.7. Both OPBC and POFA are discharged in very little amounts, typically less than 2m^3 daily for a mill having FFB crushing capacity of 45tons/hour. OPBC is a very hard lump and is suitable as an aggregate for structural lightweight concrete (Figures 1.8 to 1.10). POFA through further treatment in research has shown wonderful performance as a supplementary cementitious material (SCM) for partial cement replacement in concrete making. The research using POFA is being investigated for Geopolymer concrete. This POFA also can be used in revolutionary 3D printing concrete for the construction industry.

Figure 1.7. Little heap of disposed POFA and OPBC at an oil palm mill [courtesy, author].

Figure 1.8. Small heap of disposed OPBC at an oil palm mill [5, permission sought].

Figure 1.9. Disposed OPBC with big lumps [5, permission sought].

Figure 1.10. Big lump of OPBC crushed into suitable aggregate sizes [5, permission sought].

1.2.5. Empty Fruit Bunch

The wood industry plays an important role in generating revenue, employment and dynamism especially in the construction and furniture industries. Construction, furniture, packing etc. depend on wood and wood-based products such as: plywood and particle panel (also called particle board, or chipboard, or fiber board).

Natural wood (solid wood) has limitations in engineering properties and dimension. The natural wood is prone to warping and splitting with humidity changes. Wood-based products such as plywood, particle panel etc. have a high market demand because of several advantages. The main raw materials for making both plywood and particle panel are from natural wood which is being depleted worldwide. Due to the increasing cost of natural wood and the continuous depletion of natural wood resources, the use of solid wastes is an alternative resource to meet the needs for the construction, furniture and packing industries.

The OPS and EFB are discharged in large quantities from oil palm mills (Tables 1.2 and 1.3, Figures 1.3 and 1.11). Both OPS and EFB solid wastes, once properly processed show high potential as alternative materials for the construction, furniture and packing industries, and can meet their specifications. Therefore, the artificial plank describes the plank using renewable and sustainable solid wastes particularly EFB and OPS for use in the construction industry, furniture industry and packing industry.

Table 1.3. Estimation of EFB discharge from one oil palm mill and use of EFB fibers in plank production (courtesy, author)

Description	Quantity
Assumed hourly crushing capacity with Fresh Fruit Bunch (FFB) of one palm oil mill	45 tons/hour
Daily duration of operation	16 hours/day
Daily total crushing capacity with FFB (45tons/hour×16 hours=720 tons)	720 tons/day
Empty Fruit Bunch (EFB) production from 100 tons FFB in wet condition	12.3 tons wet EFB
Daily total EFB production [(12.3 tons EFB/100 tons FFB) ×720 tons FFB=88.56 tons] in wet condition	88.56 tons wet EFB/day
EFB fiber in dry condition/day/mill = [(1/3) ×88.56 tons/day =29.52 tons]	**29.52 tons dry EFB fibers /day/mill**
Annual EFB fiber production in dry condition/mill = 29.52tons ×365days = 10,775×70% optimization (assumed)=7,543 tons	**Dry EFB fibers=7,543 tons/year/mill**
Plank size = 12mm thick ×1.2m wide×2.4m long (similar to 12mm thick Plywood) Plank density = 890-1,090 kg/m^3 (12mm thick Plywood density= 700 kg/m^3)	
Required dry EFB fiber for one plank	25 kg/plank
Daily plank production using 29.52 tons dry EFB fibers (29.52 tons×1,000kg=29,520 kg÷25 kg/plank = 1,180 units plank)	1,180 units plank/day/mill
Annual Plank production =1180×365 days=430,700 units×70% optimization (assumed)=301,490= 300,000 (rounded) units **(Note Plank size = 12mm thick ×1.2m wide×2.4m long)**	**300,000 units planks/year/mill**

OPS being strong and porous improves the sound insulation properties in partition walls or ceilings. Having very low aggregate crushing values (ACV) and aggregate impact values (AIV), OPS aggregate shows very good energy absorption properties. OPS is superior quality than thin wood sheet, wood chips and sawdust. Fiber obtained from EFB is strong in tensile strength. Figure 1.12 shows a single EFB which is full of fibers. The shredded EFB fibres are required pre-treatment to remove oil coating and other impurities. Thus, the pre-treatment either using detergent mixed boiled water and Urea solution can be employed to prepare EFB fibers as light-golden or off-white as shown in Figure 1.13. Therefore, the plank (Figures 1.14 and 1.15) if made with combination of EFB fiber and OPS will enhance density, strength, sound and heat insulation, aesthetics etc. This kind of high-density plank can be used for many purposes such as construction, furniture, packing etc.

Figure 1.11. Large heap of disposed EFB at an oil palm mill [courtesy, author].

Figure 1.12. Empty Fruit Bunch (EFB) full with fibers [courtesy, author].

a. Light-golden colour of EFB fibers after treatment with detergent mixed boiled water

b. Off-white colour of EFB fibers after treatment with Urea solution

Figure 1.13. Treated EFB fibers in pleasant colours suitable for making any product [courtesy, author].

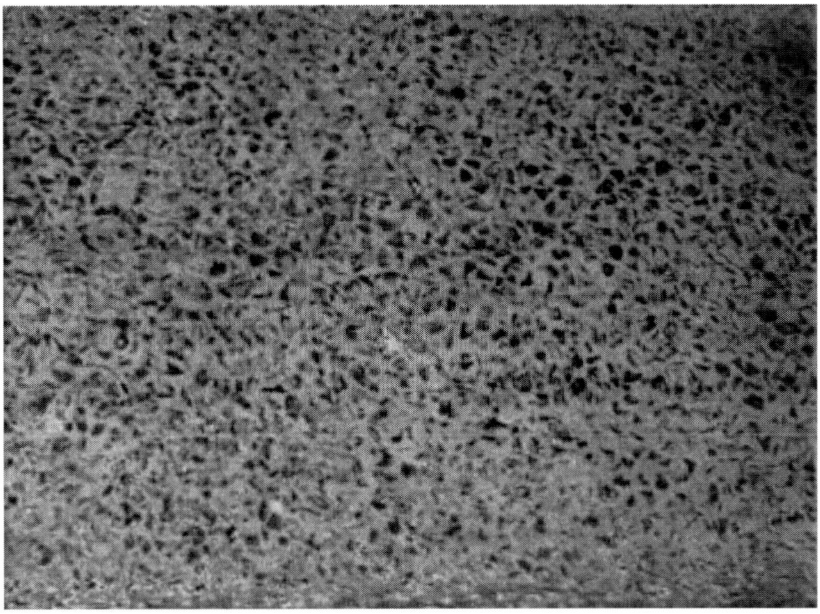

Figure 1.14. Aesthetic surface appearance of high-density plank made with OPS and EFB fibers (courtesy, author].

Figure 1.15. Verifying the strength capacity of high-density plank (density of 1,090 kg/m^3) made with OPS and EFB fibres (courtesy, author).

1.3. Evaluation Criteria for Solid Waste as Renewable Material

Generally in concrete making, ingredient must fall either under the binder (like cement and other supplementary cementitious materials) group or filler group (fine and coarse aggregates forming body of structure). Normally, binder occupies about 20% volume in concrete and filler occupies about 70% volume. The very common disposed solid wastes during oil extraction at oil palm mills are OPS, EFB, nut fibers, OPBC and POFA. Usually, the disposed OPS, EFB and nut fibers are in very large quantities. The nut fibers and OPS are used as fuel for boilers in mills. The resulting residues from boilers are OPBC and POFA. OPS solid waste, being identical with conventional 10mm size of coarse aggregate is identified as filler (aggregate) for concrete making and, both OPS and EFB solid wastes as main ingredients are for artificial plank production. OPBC being very hard acts as aggregate for lightweight concrete [5]. POFA with pre-treatment is identified as a wonderful supplementary cementitious material (SCM) for partial cement replacement in concrete. The concrete made using 330 kg POFA having median particle size (d_{50}) of 2.06 micron and 220kg cement has shown the 28-day compressive strength of 95N/mm^2 [6]. Asrah et al. (2015) [7] have reduced the POFA median particle size (d_{50}) to 1.19 micron for better engineering performance in cement paste. However, the optimum benefit can be achieved if solid wastes particularly disposed from oil palm mills as renewable resources meet the following requirements [4, 8]:

1.3.1. Locally Available in Large Quantity

The construction industry needs large amounts of concrete hence, concrete producing ingredients must be readily available in the local area. For example, a low-cost single-story house of 58.68m^2 floor area needs about 17m^3 concrete. Hence, disposed OPS solid waste at each oil palm mill is a large amount as seen in Table 1.2 and Figure 1.3, and thus OPS can be used as an aggregate in concrete production for the local construction projects like plantation housing, rural housing, non-structural components etc. Similarly, a particle board mill can be set up next to an oil palm mill to produce artificial planks using OPS and EFB for meeting the market demand. Therefore, OPS, EFB, OPBC and POFA are considered as renewable resources as they are discharged from oil palm mills daily.

1.3.2. Consistent and Low Unit Cost Resulting Cost-Efficiency

It is noted that the same solid waste produced in various mills may show variable properties. The solid wastes should be analyzed thoroughly to find out the representative performance before use. Therefore, it is recommended to use solid wastes having less deviation in properties as construction material.

The price of raw materials at origin and transport cost to bring raw materials to the mill will influence the overall cost. The finished product can be transported by road or water way. Both OPS and EFB being solid wastes can be considered as very low-priced raw materials at origin. Thus it is important to identify the strategic location for setting up the mill if artificial plank production using such solid wastes is considered. It is noticed that the physical and mechanical properties of OPS aggregate are consistent as investigated by many researchers (Table 2.1, Chapter 2). The use of low-cost raw materials in concrete or artificial plank production can reduce the total cost significantly. The overall gain using solid waste depends on economic, social and environmental aspects.

1.3.3. No Processing Cost or Minimum Processing Cost

It is another important agenda whether the raw material needs any processing or pre-treatment before mixing in the final product. During nut extraction at oil palm mills, OPS is discharged (Figure 1.3) as broken shells thus it does not need any extra crushing exercise. It may need a little pre-treatment to remove oil coating etc. if fresh OPS is selected to use. Fresh OPS is traced with little oil coating and this can be removed, as several treatment methods are reported in literature. The simplest way to remove such oil coating is to keep OPS outside the shade. The author in his research has kept OPS outside the shade for weathering for a few months. The weathered OPS is shown in Figure 2.1 (Chapter 2).

EFB needs processing for separating fibers from the bunch and, the separation and pre-treatment should be done to remove the oil coating and other impurities (Figure 1.13). Pre-treated dry fibers must be kept in a dry place protected from the moisture. OPBC lump needs a crushing exercise to bring down the required aggregate size. Researchers are focusing on POFA treatment through heat and grinding in order to bring the particles down to ultra-micron size.

1.3.4. No Health Hazard and Easy to Handle

In the non-hazardous category, the main harmful characters are corrosivity, ignitability, reactivity and toxicity. Both OPS and EFB materials are organic and combustible. Ignitability of OPS aggregate can be minimized once it is used in concrete. Being organic in nature it will not contaminate or leach to produce hazardous toxic waste when it is cast in concrete [9]. OPS aggregate through chemical properties is considered harmless aggregate (Table 2.4, Chapter 2) and does not pose any health risk during handling as a raw ingredient. More information is shown in Section 8.3 of Chapter 8 and elsewhere while making a low cost house using OPS concrete manually. EFB being an agriculture product does not show any health risk in handling. Both OPBC and POFA can be considered as casual waste like fly ash. They are handled easily at oil palm mills as well as in laboratory investigations.

1.3.5. Improvability and No Degradation on Finished Product

Improvability of any non-hazardous solid waste is explained that it has a high potential to improve the workability during construction. It must show a better relation with other conventional materials and the final products are exhibited with overall improvement including high volume stability. More investigations using solid waste should be done to meet the requirements. In all stages, the decisions on the use of such solid waste is to be made based on laboratory investigation including specifications and practical aspects.

Both OPS and EFB are organic in nature and are degradable materials. The degradation will not occur once OPS is bound in concrete (Figure 4.3, Chapter 4; Figure 5.3, Chapter 5). The presence of air and moisture are required for decay to happen. Particularly, it is essential to verify the performance of solid waste while using in concrete on:

i. Fresh concrete property and
ii. Hardened concrete properties for strength and durability

The footbridge built over a main drain about 15 years ago using OPS concrete is still in service (Section 8.2, Chapter 8). Generally, the durability of concrete mostly depends on the quality of the mortar phase particularly cement paste in concrete as it controls the moisture movement and other harmful agents. Further details on the concrete's durability are explained in Section 5.11, Chapter 5.

Artificial planks using both OPS and EFB are non-structural components like plywood and particle board for use in the construction industry, furniture industry, packing industry etc. so durability is not very important issue.

Both OPBC and POFA being burnt products as inorganic materials are discharged from boilers, so they do not show any degradation on resultant concrete.

1.4. Chapter Remarks

This chapter has described the amount of solid wastes discharged from oil palm industry and the potential utilization of these solid wastes as renewable materials in relation to the construction industry, furniture industry etc. It has also explained how solid wastes are selected as per need. However, the main focus of this book is the use of OPS solid waste as an aggregate in lightweight concrete and performance of structures made with such lightweight concrete. They are elaborated in the next chapters.

References

[1] *Oil Palm in Malaysia-E-tawau*, www.etawau.com/OilPalm/OilPalm.htm. Retrieved on 25 Oct 2016.

[2] R. J. Spence and D. J. Cook. 1983. *Building Materials in Developing Countries.* John Willy & Sons Publications.

[3] Y. Shirai, M. Wakisaka, S. Yacob, M. A. Hassan and S. Suzuki. 2003. Reduction of Methane Released from Palm Oil Mill Lagoon in Malaysia and its Countermeasures, *Mitigation and Adaptation Strategies for Global Change* (8):237 – 252.

[4] M. A. Mannan. 2001. *A study on the use of Oil Palm Shell in Lightweight Concrete.* PhD. Thesis. Universiti Malaysia Sabah.

[5] M. A. Mannan and K. Neglo. 2010. Mix design for oil-palm-boiler clinker (OPBC) Concrete. *Journal of Science and Technology.* **30** (1). pp 111-118.

[6] M. A. Megat Johari, A. M. Zeyad, N. Muhamad Bunnori and K. S. Ariffin. 2012. Engineering and transport properties of high-strength green concrete containing high volume of ultrafine palm oil fuel ash. *Construction and Building Materials*. **30**. pp 281-288.

[7] H. Asrah, A. K. Mirasa and M. A. Mannan. 2015. The performance of ultrafine palm oil fuel ash in suppressing the Alkali silica reaction in mortar bar. *International Journal of Engineering and Applied Science*. **2**(9). pp 60-66.

[8] Nontananandh Supakij. 1990. *Industrial waste utilization as construction materials by chemical stabilisation*. Ph. D Thesis. Dept. of Civil Engineering, Kyoto University, Japan.

[9] Shakor R. Badaruddin. 1993. Use of waste materials in construction- a review. *Civil Engineering Journal*. Universiti Teknologi Malaysia. **6**. pp. 40-49.

Chapter 2

PROPERTIES OF OIL PALM SHELL LIGHTWEIGHT-AGGREGATE AND PRE-TREATMENT

Keywords: Oil Palm Shell (OPS), lightweight aggregate, engineering properties, pre-treatment

2.1. INTRODUCTION

The engineering properties such as physical, mechanical and chemical of aggregate as construction material are main characteristics to determine whether it is suitable for use. Thus, the suitability of Oil Palm Shell (OPS) aggregate in lightweight concrete production has been investigated since the 1970's with initiation at the University of Benin, Benin, Nigeria [1] The research using OPS as coarse aggregate in concrete making is mostly conducted in Malaysia, Nigeria, Togo and Ghana where oil palm plantation is available. Therefore, this chapter elaborates the physical, mechanical and chemical properties of OPS aggregate in order to satisfy the requirement as lightweight coarse aggregate to produce the particularly structural lightweight concrete.

Lightweight aggregates can be grouped as natural aggregates, artificial aggregates, aggregates from industrial by-products and aggregates from organic wastes. Most artificial lightweight aggregates are manufactured in western countries with their own technology. The suitability of technology for producing artificial lightweight concrete is well documented and proven. The use of solid wastes, particularly OPS organic solid waste as coarse aggregate can potentially reduce the cost of lightweight concrete.

The physical properties of OPS are shape, size, thickness, particle size analysis, specific gravity, density and water absorption. The mechanical properties investigated such as aggregate impact value (AIV), aggregate crushing value (ACV) and Los Angeles abrasion value (AAV) of OPS aggregate are very significant compared with conventional coarse aggregates. The chemical properties of OPS aggregate indicate whether OPS is harmless and is suitable in concrete making.

Several pre-treatment methods are employed for the quality improvement of OPS aggregate. They are as: (1) detergent, (2) borate solution, (3) sodium dichromate solution, (4)

ferrous sulphate solution, (5) cupric sulphate pentahydrate solution, (6) acetic acid solution, (7) polyvinyl alcohol solution, (8) heat, and (9) bitumen.

2.2. Aggregate Properties

The properties of concrete depend upon the properties of the individual ingredients and their compatibility. In concrete, typically about 9-15% of space is occupied by binder (cement), 15-16% by water, 25-35% by fine aggregate and 30-45% by course aggregate. Aggregate is a cheaper material compared with cement and, it also has high level of structural stability and better durability than cement paste alone. The properties of aggregates influence the structural performance and durability of concrete. Both fine and coarse aggregates are commonly known as filler in concrete. Aggregates as filler, form the concrete body and the aggregate particles are held together by cement paste. The ingredients and their compatibility through physical, mechanical, chemical and thermal expansion characters influence the engineering properties of concrete in fresh and hardened states. The properties of aggregates are to determine first in order to design and produce the structural and non-structural concrete. Aggregates are not branded materials like cement. The size, shape, surface texture, porosity and water absorption capacity of aggregate influences water demand in concrete mix, workability and strength. The gradation of aggregate particle size influences water demand in mix design, segregation, bleeding and strength. The aggregate density, specific gravity, strength and modulus of elasticity influence the concrete density and strength. The chemical stability of aggregate influences the durability of concrete. The surface texture of aggregates influence the bond between aggregate and cement paste. The surface textures of aggregates also influence the flexural strength more than compressive strength of concrete. The rough surface texture and large surface area of angular aggregate influence the adhesive force between aggregate and cement paste greatly [2].

Generally, the aggregate shapes are of different types such as round, irregular, cubical, flaky, angular and elongated. Flaky aggregates are defined as those whose least dimension is less than 3/5th of its mean dimension. For example, the mean dimension of an aggregate passing through a 12.5mm sieve and retained on a 4.75mm sieve is $[12.5+4.75]\div 2=8.63$mm. If the least dimension is less than $3/5 \times 8.63=5.18$mm, the aggregate is flaky. The flakiness index is the % by weight of flaky aggregates. Elongated aggregates are also defined as those whose length is 1.80 times its mean dimension. If the mean dimension as shown above is, say, 8.63mm for a length above 15.53mm ($1.80\times 8.63=15.53$mm), the aggregate is called elongated. The elongation index is the % by weight of elongated aggregates. Aggregates of flaky and elongated shapes have a larger surface area as compared with rounded or cubical aggregates of the same weight. Hence, a higher fine aggregate and w/c ratio are required to coat the surface of flaky and elongated aggregates. Concrete produced with flaky and elongated aggregates is prone to segregation. Hence, it needs extra attention in mix design to avoid segregation. Therefore, the physical, mechanical and chemical properties of OPS aggregate are discussed below in order to satisfy the requirement as lightweight aggregate to produce the structural lightweight concrete.

2.2.1. Physical Appearance of OPS as Lightweight Aggregate

Lightweight aggregates are classified as natural and artificial. Natural lightweight aggregates are obtained mostly from volcanic rocks. Artificial lightweight aggregates are produced in different countries for three main reasons:

a) Fulfilling the demand for aggregates
b) Exploiting large volume of industrial wastes, and
c) Use of different lightweight concretes for:
    1) Structural
    2) Non-structural and
    3) Insulation purposes

Table 2.1 shows the physical and mechanical properties of OPS as lightweight coarse aggregate. Generally, the maximum size of conventional coarse aggregate used in normal weight concrete is 20mm. Therefore, the maximum size, 14mm of OPS aggregate investigated by researchers is suitable in concrete making.

Aggregate particles are of different shapes such as round, irregular, flaky, angular and elongated. The shape and surface texture of aggregate influence the flexural strength of concrete more than compressive strength. Roughness, area and shape of surface of aggregate, influence greatly the adhesive force between aggregate and cement paste [2].

The size, shape and surface texture of aggregates influence the workability of concrete. The shape of conventional normal weight aggregate is influenced by both types of parent rocks and type of crusher used. It also depends on the reduction ratio. The reduction ratio is the ratio of the size of material fed into crusher to the size of final product. The conventional normal weight aggregates such as Schists, Slates and Shales produce the flaky form and, Granite, Basalts, Sandstone, Limestone and Quartzite form more or less equidimensional in size. The rounded aggregates are more preferable to angular aggregates from an economic point of view. The angular aggregates give higher strength, higher bond characteristics and greater durability than round aggregates. The excessive flaky aggregates produce very poor concrete. The void content and water demand in concrete mix are influenced by the shape and surface texture of aggregates.

The thickness of OPS depends on the species of the palm tree from which the nut is collected. Depending on the plant species, the thickness of shells ranges from 0.5-8.0mm. Dura species has the thickest shell varying from 2-8mm, whereas Tenera has 0.5-4.0mm thick shells and Pisifera is practically shell-less. In practice, the fruits of different species are processed in the crude oil palm mill. As such, shells of different thicknesses are discharged in the mill. The majority of shell thicknesses in Okafor's (1988) [3] investigation are between 2 mm and 3 mm. The shell thicknesses of OPS aggregate investigated by several researchers are in the range of 0.5-3.5mm (Table 2.1). Physically, the shells are hard and resistant to bio-deterioration [4]. The shapes of OPS mainly depend on the breaking pattern of the nut. The surface texture of OPS for both concave and convex faces is fairly smooth [3, 4]. The broken edges are rough and spiky. During oil extraction at oil palm mills, OPS, the hardened pericarp of the oil palm nut are discharged as roughly parabolic or semi-circular, flaky and other irregular shapes as shown in Figure 2.1. The shells should have sufficient strength to be used as coarse aggregate in concrete making.

Fineness modulus (F. M.) values of OPS aggregate revealed by researchers are in the range of 5.78-6.30 (Table 2.1) and are comparable to conventional coarse aggregate. The F. M. is a numerical index of fineness giving an idea on the mean size of the aggregate. The F. M. of aggregate is a prime indication for the most economic mix design to obtain the required workability and strength. The particle size analysis of OPS; conventional coarse aggregate, Granite and river sand is shown in Figure 2.2.

OPS aggregate in convex surfaces has pores of 16-24micrometer [5] which are also a contributing factor in increasing void content as well as water absorption. Table 2.2 shows water absorption capacities of some conventional lightweight aggregates in the range of 5-50% [6-8]. The water absorption capacity of OPS aggregate is 18.69-33% and is comparable with the conventional lightweight aggregates. This high water absorption will enhance internal curing once lightweight concrete is made with OPS aggregate. The pore structure and surface condition of lightweight aggregate govern the quantity of water absorption. The water absorption capacity may vary from 5 to 25% for lightweight aggregate and it may be about 2% for normal weight aggregate [6, 8]. The absorbed water of aggregates influences the bulk density, workability, thermal insulation, creep and shrinkage of the resultant lightweight concrete.

The fresh concrete properties and, bond between aggregate and mortar phase are influenced significantly by physical properties such as roughness, shape and texture of aggregates. The smoothness or roughness of the aggregate reflects the surface texture. The glassy, smooth, granular, rough, crystalline, porous and honeycombed textures are the visual properties of the surface of aggregate [9].

2.2.2. Specific Gravity and Density of OPS Aggregate

The specific gravities of OPS aggregate revealed in several studies are in the range of 1.17 - 1.34 which is less than the limit value, 2.2 for lightweight aggregate [10]. The bulk density of OPS aggregate is between 500 and 685kg/m^3 which is within the usual limit of 250 – 1,000 kg/m^3 classifying OPS as lightweight aggregate. The bulk density is a function of both the void space between the aggregate particles and the density of the aggregate particles. The voids depend on the size, grading and shape of the particles, and on the packing mode, whether they are loosely deposited, vibrated or compacted, and the moisture content in the aggregate. The typical bulk densities of conventional lightweight aggregates are shown in Table 2.3 [3, 6-8, 11]. According to British standards [12], the density of lightweight aggregate cannot exceed 960 kg/m^3 for coarse aggregate.

2.2.3. Mechanical and Chemical Properties of OPS Aggregate

The porosity and strength of aggregate particles affect the compressive strength of resultant concrete. Generally, porous aggregates show low concrete strength. The tensile strength of lightweight aggregate is an important property. Aggregates with densities of 1400 to 1500 kg/m^3, can develop maximum particle tensile strengths of 2.3 to 4.6 N/mm^2 [8].

The density, pore structure and irregular texture of aggregates define the modulus of the elasticity of aggregates. There is no standard method of test to obtain the modulus of elasticity for aggregates. The modulus of elasticity of an aggregate can be determined from the result of ultrasonic tests using single aggregate.

The moduli of elasticity of expanded clay and shale are in the range of 0.5×10^4 to 1.5×10^4 N/mm^2 [8]. The modulus of elasticity of the aggregate affects the strength of the resultant concrete. Generally, the higher the modulus of elasticity of aggregate, the higher the strength and stiffness of concrete. So far, no investigation has been performed on OPS aggregate for the modulus of elasticity. However, the mechanical properties of OPS aggregate like AIV, ACV and AAV values are lower than the limit, which are beneficial factors using OPS in concrete making (Table 2.1). Table 2.4 shows the chemical properties of OPS aggregate which is considered as harmless aggregate and is suitable for concrete making.

Table 2.1. Physical and mechanical properties of OPS aggregate [courtesy, author]

Description	Mannan 2001	Teo 2007	Ng 2010	Alengaram et al. 2011	Shafigh et al. 2013	Islam et al. 2016	Remark on OPS as lightweight coarse aggregate
Physical properties							
Maximum size (mm)	12.5	12.5	12.0	14	-	-	Use of maximum size of coarse aggregate in concrete yields: • reduction of cement, water, drying shrinkage, heat output and overall construction cost • higher volume stability
Shell thickness, mm	-	0.5-3.0	2.5-3.5	0.7-3.5	-	-	-
Specific gravity (saturated surface dry)	1.17	1.17	1.33-1.34	1.27	1.21	1.23	OPS having specific gravity of 1.17-1.34 as within the normal limit of 0.05 – 2.20 for most lightweight aggregates [10]
Water absorption for 24 hours (%)	23.30	33.0	21.30-22.63	25	20.36	18.69	Having high water absorption of OPS aggregate enhancing internal curing
Bulk density (compacted), kg/m^3	590	500-600	599-646	620	673	685	LWA limit of 250 – 1,000 kg/m^3 classifying OPS as LWA; most LWAs used in structure are in the range of 560 to 960 kg/m^3 [13]
Fineness Modulus (F.M)	6.24	6.08	6.23-6.30	6.24	5.78	5.92	-
Flakiness Index (%)	65.17	-	-	-	-	-	-
Elongation Index (%)	12.36	-	-	-	-	-	-
Mechanical properties							
Aggregate Impact value, AIV (%)	7.86	7.51	3.05-6.60	3.91	-	-	Generally, limit of AIV for: • Normal construction=45%

Table 2.1. (Continued)

Description	Mannan 2001	Teo 2007	Ng 2010	Alengaram et al. 2011	Shafigh et al. 2013	Islam et al. 2016	Remark on OPS as lightweight coarse aggregate
							• Pavement wearing surface course=30% • Heavy duty floor =25% However, low values of OPS aggregates indicating good energy absorbing capacity
Aggregate Crushing value, ACV (%)	-	8.00	2.33-7.12	-	-	-	Generally, limit of ACV for: • Road and pavement =30%. • Other structural purposes = 45%. However, low values of OPS aggregates indicating good energy absorbing capacity
Aggregate Abrasion value, (Los Angeles), AAV (%)	4.80	4.90	2.90	-	-	-	Generally, limit of AAV for: • Pavement wearing surface=30% • Other applications=50% However, low values of OPS aggregates indicating good energy absorbing capacity

Table 2.2. Water absorption capacities of conventional lightweight aggregates [6-8]

Aggregate	Water absorption for 24 hours under atmospheric pressure (% by volume)
Volcanic tuff	7 - 30
Porous limestone	5 - 30
Expanded clay	5 - 30
Expanded shale and slate	5 - 15
Sintered pulverised fuel ash (Fly ash)	20
Foamed lava	10
Foamed blastfurnace slag	10 - 35
Pumice	50

Table 2.3. Densities of conventional lightweight aggregates [6-8, 11]

Aggregate	Air-dry density (kg/m^3)
Clinker	720 - 1040
Cork granule	68.50 - 70.50
Expanded clay	300 - 900
Expanded shale and slate	400 - 1200
Foamed blastfurnace slag	400 - 1100
Foamed lava	800 - 1400
Perlite	40 - 200
Porous limestone	750 - 1300
Pulverised fuel ash (Fly ash)	600 - 1100
Pumice	350 - 750
Sintered colliery waste	500 - 1000
Vermiculite	60 - 200

Table 2.4. Chemical properties of OPS aggregate [14]

Properties	(%)	Remark on OPS as lightweight coarse aggregate
Loss on ignition at 550°C	98.5	-
Ash at 550°C	1.53	-
Total nitrogen (N)	0.41	-
Chloride (Cl^-)	0.0007	Max total of 0.05% based on [15] for reinforced concrete made with OPC. However, very low value of OPS aggregate indicating non-harmful in concrete making
Silica (SiO_2)	0.0146	Very low value of silica content in OPS suggesting alkali-silica reaction (ASR) unlikely to happen
Aluminum (Al_2O_3)	0.13	Little amount of Aluminum and Iron indicating no pozzolanic property in OPS aggregate
Iron (Fe_2O_3)	0.0333	
Calcium (CaO)	0.0765	
Magnesium (MgO)	0.0352	-
Sodium (Na_2O)	0.0016	-
Potassium (K_2O)	0.0004	-
Total carbon	-	-
Sulphur (as SO_3)	0.0008	[16] sets a limit of 1%

Figure 2.1. OPS coarse aggregate of different shapes left outside the shade for weathering [courtesy, author].

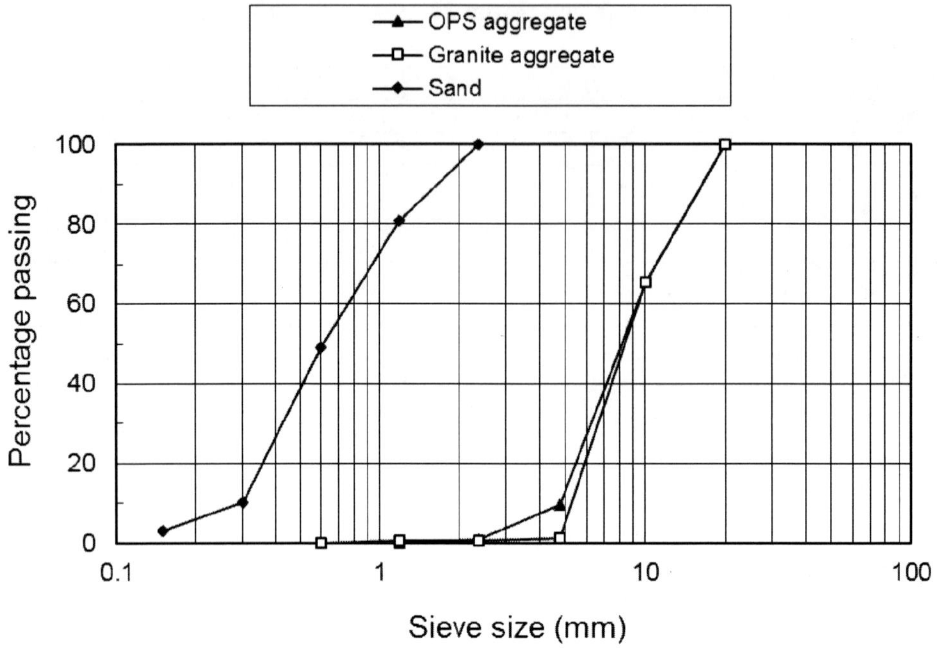

Figure 2.2. Gradation of OPS, crushed Granite and river sand [17].

2.3. Pre-Treatment on OPS Aggregate

In oil palm mills, the hard shells are directly attained by breaking the nuts with machinery. A sample of OPS aggregates is shown in Figure 2.1. Normally, OPS aggregates are composed of different shapes as shown in this figure.

Pre-treatment is necessary for OPS aggregate as it contains dust and an oil coating. Various pre-treatment processes are mentioned in the literature to remove the impurities from aggregates [4, 6, 18-21]. The processes are as:

a) Partial oxidation of organic aggregate
b) Waterproofing
c) Neutralization with alkali or precipitation of tannates, or sulphate treatment
d) Mixing with lime or calcium chloride for better performance of concrete as an accelerator
e) Micro-organism treatment in boiled water with ferrous sulphate, and
f) Removing oil coating with detergent and water

However, several pre-treatment steps have been employed on air-dried OPS aggregate to improve the quality of aggregate. They are :

1) Detergent
2) Borate solution
3) Sodium Dichromate solution
4) Ferrous Sulphate solution

5) Cupric Sulphate Pentahydrate solution
6) Acetic Acid solution
7) Polyvinyl Alcohol solution
8) Heat and
9) Asphalt

Mannan et al. (2006) [22] have investigated pre-treatment on OPS aggregate to verify the basic changes on OPS aggregate through water absorption as a physical property and aggregate crushing value (ACV) as a mechanical property. They are elaborated below.

2.3.1. Pre-Treatment with Detergent

Both Okafor et al. (1996) [21] and Mannan (2001) [17] have done pre-treatment of OPS using detergent mixed water. Freshly disposed OPS aggregates from crude oil palm mills are collected and left under a shed for air-drying [17]. These dried shells are sieved to remove the fibres, and washed in potable water mixed with detergent to remove the oil coating and other impurities. Finally, OPS aggregates are rinsed with potable water to remove the detergent. This pre-treatment method is also employed during precast footbridge construction as explained in Section 8.2 of Chapter 8.

2.3.2. Pre-Treatment with Borate Solution

Mannan et al. (2006) [22] have used full concentrated borate solution for pre-treatment of air dried OPS aggregate. They have checked the water absorption and aggregate crushing value (ACV) using this borate treated OPS and the findings are compared with OPS aggregate having no pre-treatment. The untreated air-dried OPS aggregate has shown water absorption and ACV of 23.32% and 7.86% respectively. Similarly, borate treated OPS aggregate has shown 20.83% and 6.90% which are considered very minor changes. OPS concrete made with this pre-treated OPS aggregate has shown the 28-day compressive strength of 21.50N/mm^2 which is 8.90% lower than OPS concrete made with OPS aggregate having no pre-treatment. The white spots are noticed on the surface of this hardened concrete as the OPS is pre-treated with borate solution. Therefore, the pre-treatment using borate solution is not recommended for OPS aggregate.

2.3.3. Pre-Treatment with Sodium Dichromate Solution

The 5, 10 and 20% sodium dichromate solutions have been used as another step of pre-treatment for air-dried OPS aggregate [22]. The sodium dichromate solution treated OPS aggregate is verified for water absorption and ACV. The water absorption and ACV values are found in the range of 20.03-20.58% and 5.97-6.37% respectively. The higher concentration of sodium dichromate solution has shown lower values for both measurements. The performance is still not significant compared with OPS aggregate having no pre-treatment. OPS concrete made with this pre-treated OPS aggregate has shown the 28-day compressive strength about 20.89-24.33N/mm^2. The yellow stains noticed on the surface of this hardened concrete, as the OPS is pre-treated with sodium dichromate solution. Therefore, such pre-treatment using sodium dichromate solution is not recommended for OPS aggregate.

2.3.4. Pre-Treatment with Ferrous Sulphate solution

The ferrous sulphate of 5, 10 and 20% solution has been used for pre-treatment of air-dried OPS aggregate [22]. The pre-treated OPS aggregate with different concentrated ferrous sulphate solutions has shown water absorption of 20.90-21.10% and ACV of 5.91-6.47%, which are also considered not significant changes. Like sodium dichromate solution, the higher concentration of ferrous sulphate solution shows lower values for both measurements. OPS concrete made with this pre-treated OPS aggregate shows the 28-day compressive strength of 19.67-20.39N/mm^2. The higher the concentration, the lower the compressive strength. The maximum strength reduction of about 16.7% is noticed with OPS aggregate pre-treated with 20% solution. This reduction is possibly due to a sulphate attack as the OPS is pre-treated with ferrous sulphate solution. The light green dust is observed on the concrete surface indicating excess use of this solution. Therefore, the pre-treatment using ferrous sulphate solution is not recommended for OPS aggregate.

2.3.5. Pre-Treatment with Cupric Sulphate Pentahydrate Solution

Mannan et al. (2006) [22] have used cupric sulphate pentahydrate solution of 5, 10 and 20% for pre-treatment of air-dried OPS aggregate. The pre-treated OPS aggregate with different concentrations of this solution have shown about 20.05-20.38% water absorption and 6.03-6.57% of ACV and this pre-treatment is not significant as compared with untreated OPS aggregate. OPS concrete made with this pre-treated OPS aggregate has shown the 28-day compressive strength of 13.38-19.07N/mm^2. The higher the concentration, the lower the compressive strength. The maximum strength reduction of about 43.3% is noticed using a 20% solution. Again, this reduction is possibly due to a sulphate attack as the OPS is pre-treated with cupric sulphate pentahydrate solution. The marine blue spots as the most significant appearance, are noticed on the concrete surface. Therefore, such pre-treatment using cupric sulphate pentahydrate solution is not recommended for OPS aggregate.

2.3.6. Pre-Treatment with Acetic Acid Solution

The acetic acid of 5 and 10% solution has been tried for pre-treatment of air-dried OPS aggregate [22]. The pre-treated OPS aggregate with different acetic acid solutions has been checked for water absorption and ACV, and are found to show no significant changes on OPS aggregate. OPS concrete made with this pre-treated OPS aggregate has shown the 28-day compressive strength of 18.82N/mm^2. The higher the concentration, the lower the compressive strength. The maximum strength reduction of about 20.3% is noticed with OPS aggregate pre-treated with a 10% solution. No physical change is observed on the concrete surface with this acetic acid pre-treated OPS aggregate. Therefore, such pre-treatment using acetic acid solution is not recommended for OPS aggregate.

2.3.7. Pre-Treatment with Polyvinyl Alcohol Solution

The polyvinyl alcohol (PVA) solutions of 10 and 20% have been used for pre-treatment of air dried OPS aggregate [22]. The PVA is a granular material and white in colour. The 20% PVA solution is done using 200gm PVA granules in a cylinder and adding distilled water to make 1,000ml. This mix is boiled at 100^{0}C until all granules are dissolved. The air-dried OPS aggregate as shown in Figure 2.3 of 3kg is kept in a mixer machine (HOBAHT) and then a 500ml PVA solution is added gradually. Mixing is done uniformly for up to 3 minutes, then 1kg of sieved and washed river sand (passing ASTM 30 sieve and retained ASTM 50 sieve) is mixed thoroughly in the mixing machine for another 1 minute for sand coating onto the surface of OPS aggregate. After mixing, the sample is kept on a plastic sheet in the laboratory for air-drying (Figures 2.4 and 2.5).

The pre-treated OPS aggregate with 10 and 20% PVA solutions has shown 4.25-7.19% water absorption and 1.04-2.09% of ACV. This pre-treatment method among all methods explained above has shown the most significant improvement as compared with untreated OPS aggregate. Again, OPS aggregates pre-treated with 20% PVA solution are exposed to 1M NaOH, 0.1M H_2SO_4 and 5.2% $MgSO_4$ solutions as aggressive conditions. The concrete itself is an alkaline product and concrete can be exposed to aggressive acidic and sulphate conditions. Hence, these three aggressive conditions are used to verify the performance of OPS as aggregate. The water absorption, specific gravity and ACV changes on 20% PVA solution treated OPS aggregate and OPS aggregate having no pre-treatment under such aggressive conditions are shown in Table 2.5. The OPS aggregate pre-treated with 20% PVA solution has shown quite impressive performance under these three aggressive conditions and is suitable to be used in concrete making.

Table 2.5. Water absorption, specific gravity and ACV properties of OPS aggregate in aggressive conditions [22, with modification]

Property investigation	OPS treatment type	Aggressive conditions		
		1M NaOH	0.1M H_2SO_4	5.2% $MgSO_4$
Water absorption (%)	No pre-treatment	35.39	26.73	27.06
	20% PVA solution	11.62	8.06	10.03
Aggregate Sp. Gravity	No pre-treatment	1.08	1.12	1.10
	20% PVA solution	1.13	1.16	1.20
ACV	No pre-treatment	8.87	9.13	8.65
	20% PVA solution	5.00	4.60	2.02

OPS concrete made with 10 and 20% PVA pre-treated OPS aggregate has shown the 28-day compressive strength of 28.22 and 32.84N/mm^2 respectively. The higher the PVA concentration, the higher the compressive strength and the maximum strength gain of about 39.2% is noticed using 20% PVA solution. The concrete failure surface indicates a better bond between PVA treated OPS aggregate and mortar, increasing the compressive strength.

No physical effect is seen on the concrete surface made with this pre-treated OPS aggregate. The investigation using PVA solution as pre-treatment for OPS aggregate and concrete made with such treated OPS aggregate has shown encouraging engineering performance [23, 24].

Figure 2.3. Air-dried OPS aggregate ready for mixing with PVA solution [courtesy, author].

Figure 2.4. PVA solution mixed OPS aggregate with glossy surface prior to sand coating onto OPS surface [courtesy, author].

Figure 2.5. Ready OPS aggregate pre-treated with PVA solution and sand coating for aggregate surface roughening [courtesy, author].

2.3.8. Pre-Treatment with Heat

Yew at al. 2014 [25] have investigated OPS aggregate through heat treatment. They have used two temperature ranges such as 60 and 150^{0}C with the duration of 30 minutes and 1 hour for pre-treatment of OPS aggregate. The concrete made with such heat treated OPS aggregate has shown the reduction of compressive strength at all ages from 1-day to 90-days. Therefore, such heat treatment is not recommended for OPS aggregate.

Figure 2.6. OPS aggregate being pre-treated with grade 80/100 bitumen [courtesy, author].

2.3.9. Pre-Treatment with Bitumen

The author has proposed another pre-treatment using bitumen mixed OPS aggregate. The grade 80/100 bitumen is heated like hot mixed asphalt (HMA) and air-dried OPS aggregate is mixed first thoroughly and finally air-dried sand (as shown in Sec 2.3.7) is mixed to form a roughened surface of OPS aggregate. The conceptual steps are elaborated in Figures 2.6 and 2.7. The bitumen can be a cost effective pre-treatment as it needs very little dosage.

Figure 2.7. Pre-treated OPS aggregate with grade 80/100 bitumen and sand coating for aggregate surface roughening [courtesy, author].

2.4. Chapter Remarks

On the basis of review, it is seen that OPS is a prospective coarse aggregate in structural lightweight concrete. It compares favourably with other materials used as lightweight aggregates. Therefore, OPS can be offered as a good substitute for coarse aggregate in lightweight concrete useful for structural purpose.

References

[1] M. O. Uviasah. 1976. *Properties of Palm kernel shells and Fibre as aggregates for making concrete*, Unpublished Bachelor of Engineering Thesis, Department of Civil Engineering, University of Benin, Benin, Nigeria.
[2] A. M. Neville. 1981. *Properties of Concrete*, 3rd edition. Pitman.
[3] Okafor Fidelis O. 1988. Palm kernel shell as a lightweight aggregate for concrete. *Cement and Concrete Research*. **18.** pp. 901-910.

[4] S. A. Salam. 1982. Lightweight concrete made from palm oil shell aggregates and rice husk. *Proceedings of A Regional Seminar*. Universiti Putra Malaysia. Sept. 15-17. pp. 177-196.

[5] U. J. Alengaram, H. Mahmud, M. Z. Jumaat. 2011. Enhancement and prediction of modulus of elasticity of palm kernel shell concrete. *Materials and Design* **32**, pp. 2143-2148.

[6] B. H. Spratt. 1974. *The structural use of lightweight aggregate concrete.* Cement and Concrete Association.

[7] M. A. Aziz, S. D. Ramaswamy, C. K. Murthy. 1978. Lightweight concrete using cork granules. *Proceedings of the International Conference on Materials of Construction for Developing Countries.* Bangkok, Thailand. August 22-24, pp 181-189.

[8] FIP. 1983. *FIP Manual of Lightweight Aggregate Concrete*, 2nd edition. London. Surry University Press.

[9] M. L. Gambhir. 2004. *Concrete Technology*, 3rd edition. India. Tata McGraw-Hill Publishing Company limited.

[10] S. Popovics. 1992. *Concrete Materials – Properties, Specifications and Testing*. (2nd edition). New York: William Andrew Publishing.

[11] R. N. Swamy. 1986. *Natural fibre reinforced cement and concrete*. Blackie and Son Ltd. London.

[12] BS3681. Methods for sampling and testing of lightweight aggregates for concrete. Part 2. *British Standard Institution*. London.

[13] J. E. Kogel, N. C. Trivedi, J. M. Barker, S. T. Krukowski. 2006. *Industrial Minerals & Rocks: Commodities, Markets, and Uses*. (7th edition). Society for Mining, Metallurgy, and Exploration Inc. (SME).

[14] C. H. Ng. 2010, *Studies on precast reinforced concrete floor panels using Oil Palm Shell aggregate*, PhD. Thesis. Universiti Malaysia Sabah, Sabah, Malaysia.

[15] BS 882: 1992. *Specification for Aggregates from Natural Sources for Concrete*. London: British Standards Institution.

[16] BS 3797. *Specification for Lightweight Aggregates for Masonry Units and Structural Concrete*. London: British Standards Institution.

[17] M. A. Mannan. 2001. *A study on the use of Oil Palm Shell in Lightweight Concrete*. PhD. Thesis. Universiti Malaysia Sabah.

[18] Hussin M. W & Zakaria F.1990. Composite materials made from vegetable fibres as agglomerated irregular micro-reinforcement and Portland cement used in pieces for low-cost housing. Vegetable Plants and Their Fibres as Building Materials, *Proceedings of the 2nd International Symposium*. Salvador, Bahia, Brazil. Sept. 17-21. pp. 77-86.

[19] S. Mindess, J. F. Young. 1981. *Concrete*. Englewood Cliffs, New Jersey. Prentice-Hall Inc.

[20] A. Short, W. Kinniburgh. 1978. *Lightweight Concrete*, 3rd edition. London. Applied Science Publishers Ltd.

[21] Okafor Fidelis O, Eze-Uzomaka O.J & Egbuniwe N. 1996. The structural properties and optimum mix proportions of palmnut fibre-reinforced mortar composite. *Cement and Concrete Research*. **26**. pp. 1045-1055.

[22] M. A Mannan, J. Alexander, C. Ganapathy, D. C. L. Teo. 2006. Quality improvement of oil palm shell (OPS) as coarse aggregate in lightweight concrete. *Building and Environment*. **41**. pp. 1239-1242.

[23] C. C. Thong, D. C. L. Teo, C. K. Ng. 2015. The effect of polyvinyl alcohol (PVA) on the mix proportion of oil palm shell (OPS) concrete. *Applied Mechanics and Materials*. **695**. pp. 297-300.

[24] C. C. Thong, D. C. L. Teo, C. K. Ng. 2016. Application of polyvinyl alcohol (PVA) in cement-based composite materials: A review of its engineering properties and microstructure behaviour. *Construction and Building Materials.* 107. pp. 172-180.

[25] M. K. Yew, H. B. Mahmud, B. C. Ang, M. C. Yew. 2014. Effects of heat treatment on oil palm shell coarse aggregates for high strength lightweight concrete. Materials and Design. 54. pp. 702-707.

Chapter 3

RESISTANCE CAPACITY OF OIL PALM SHELL IN ACID, SALT AND ALKALI SOLUTIONS

Keywords: Oil Palm Shell (OPS) aggregate, wetting and drying cycle, aggressive solutions, water absorption, Aggregate impact value (AIV), Aggregate crushing value (ACV)

3.1. INTRODUCTION

Concrete, being man-made rock is the most used material in the construction industry worldwide. General ingredients of concrete are cementitious material, aggregates and water. Aggregates, both fine and coarse are commonly considered filler and they make the body of any concrete structure. They occupy about 60 to 80% of the volume and 70 to 80% of the weight of concrete.

Due to rapid urbanization and to meet the construction demand, natural stones being inorganic, and of both high and marginal quality are being depleted. Thus researchers are focusing on renewable resources as alternative construction materials. As such, large amounts of oil palm shell (OPS) as renewable organic solid waste is discharged at oil palm mills. This OPS is also regarded as green aggregate as it does not need extra energy to be produced in required aggregate sizes unlike crushed stone produced at quarries. This large amount of OPS can contribute to overcoming the over-dependence on depletable resources for concrete production. This is especially true for low cost house, plantation house, non-structural precast components such as Industrialized Building Systems, IBS products for drains, retaining structures etc. The use of solid waste as an ingredient in concrete is not a new idea. Since civil engineering projects are of large-scale and need a great deal of material, the major technical characteristics of waste materials must be primarily understood before each particular use.

Agricultural waste is considered as useless material, which has no other practical value. Due to the development of technology, many agricultural wastes can now be transformed into a form, which as renewable resources, are recoverable, replaceable and saleable. The use of solid waste or by-products in the construction industry can reduce the total cost significantly.

It has been found that the bulk density of OPS aggregate is about 590 kg/m^3 and resulting density hardened OPS concrete (Figure 3.1) is less than 2000 kg/m^3 [1], which classifies them as lightweight [2]. The use of OPS aggregate as coarse aggregate in construction is still relatively new and has yet to be widely utilized in the construction industry. A number of researches [3, 34] have been carried out to study both OPS aggregates and OPS concrete properties.

In recent decades, acid rain occurs due to industrialization activities such as automobile exhaust, industrial emissions, coal combustion etc. It has a widespread impact on the environment especially on construction facilities made with concrete. In the Guangdong Province of China, 50% of storms contain acid rain and the annual pH value of precipitation is less than 4.5 [35]. The acid rain fall covers at least one third of China as reported.

Concrete is high alkaline man-made rock having a pH value in the range from 12 to 13.5. Upon exposure to the acid rain situation, physical and chemical reactions occur resulting in a reduction of the concrete's pH value. The reactions show the dissolution and decomposition of cement hydration products. Hence, concrete shows cracking, losses of strength and weight, and ultimate structural failure.

Salt containing chlorides such as sodium chloride, calcium chloride etc damage the concrete. Most salts are in the range from neutral to mildly acidic and it is hygroscopic in nature to attract and retain water. Chloride plays a major role on corrosion in reinforced concrete.

The durability of a concrete structure is affected by several factors such as constituent materials, process and aggressiveness in the environment. Thus it is an important agenda to verify the performance of the individual concrete making ingredients through physical, mechanical and chemical conditions.

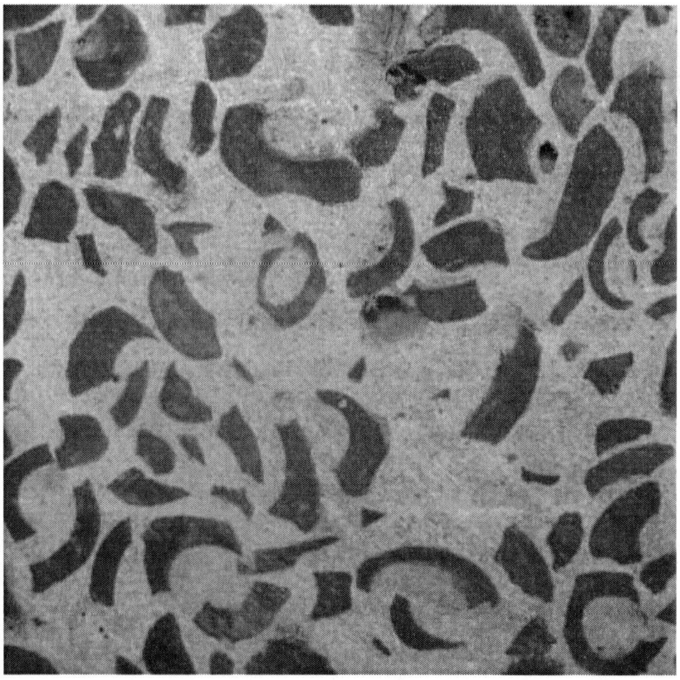

Figure 3.1. Insight of OPS concrete [black marks indication for OPS as coarse aggregate, [36 with permission].

3.2. External Adverse Effects on Concrete Durability

The durability for building materials may be defined as the ability to resist adverse external influences such as climatic conditions, fire, chemical attack and mechanical damage. The principal conditions for breakdown mechanisms in concrete are access to water and the permeability of the structure which defines how fast aggressive liquids or ions may be transported into or through the material.

Aggressive ground-water, sulphate, polluted air and spillage of reactive liquids in situations such as sewers, sewage treatment facilities and buildings in natural hot-spring areas cause chemical attacks on concrete. The chemical attack is mostly on cement paste and usually the aggregates are not susceptible to chemical attacks.

Carbonation is the reaction between carbon dioxide and the minerals in the cement paste. Carbon dioxide is available in the atmosphere and in moisture. Most lightweight aggregate particles are more porous than normal weight aggregate particles, and allow more diffusion of gases such as carbon dioxide. Carbonation promotes the corrosion of embedded reinforcement in concrete. In structural lightweight concrete, the possibility of an accelerated carbonation is due to the diffusion of carbon dioxide through pores of aggregates.

3.3. Water Absorption Properties of Different Lightweight Aggregates

The pore structure and surface condition of lightweight aggregate govern the quantity of water absorption. The water absorption capacity may vary from 5 to 25% for lightweight aggregate and it may be about 2% for dense aggregate [37, 38]. Typical water absorption capacities of inorganic lightweight aggregates are shown in Table 2.2 (Chapter 2). The water absorption of OPS organic aggregate has been investigated and it is found to be in the range from 18.69 to 33% [9, 13, 15, 26-34]. Usually, the absorbed water of aggregates influences the bulk density, workability, thermal insulation, creep and shrinkage, and acts as a reservoir for internal curing of the resultant lightweight concrete.

3.4. Materials and Test Methods

OPS as coarse aggregate has been investigated, treated with aggressive solutions such as acid, salt and alkali through water absorption, AIV and ACV tests. The chemicals used in the tests which act as deleterious solutions were salt, acid and alkaline solutions of 3% sodium chloride (NaCl), 1% hydrochloric acid (HCl) and sodium hydroxide (NaOH) of pH 13 respectively. Potable water was used as a control condition.

OPS aggregates were exposed to wetting and drying cycles under various solutions. One cycle consists of three days of wetting and four days of drying under laboratory conditions with an ambient temperature of 22-28°C and relative humidity of 74-88%. Specimens were subjected to 20 cycles of wetting and drying.

OPS aggregates were treated in different solutions as shown in Table 3.2 [39]. After exposure, the properties of OPS aggregates tested, included (i) water absorption, (ii) aggregate impact value (AIV) and (iii) aggregate crushing value (ACV).

Table 3.2. Exposure conditions of OPS aggregate [39]

Exposure condition of OPS aggregate	Exposure Duration
a. Potable water b. 3% NaCl solution c. 1% HCl solution d. NaOH of pH 13 solution	20 cycles [one cycle with 3 days immersion followed by 4 days air dried at laboratory condition]

AIV and ACV tests were carried out with minor modifications [40, 41]. BS has mentioned the aggregate passing the 12.5mm size and retained on a 10mm sieve is required to test. Based on surveys at oil palm mills, more than 80% OPS is less than 10mm. Therefore, OPS aggregate retained on a 4.75mm sieve was taken for these tests.

3.5. Water Absorption of OPS Aggregate Treated in Water, HCl, NaCl and NaOH Solutions

The water absorption of OPS aggregate initially was high, up to 22.63%. The high water absorption of OPS is due to its natural biological condition where it is porous and not as dense as rock aggregates. OPS aggregate has a large number of internal pores and this causes the porosity of the aggregate to be considerably high.

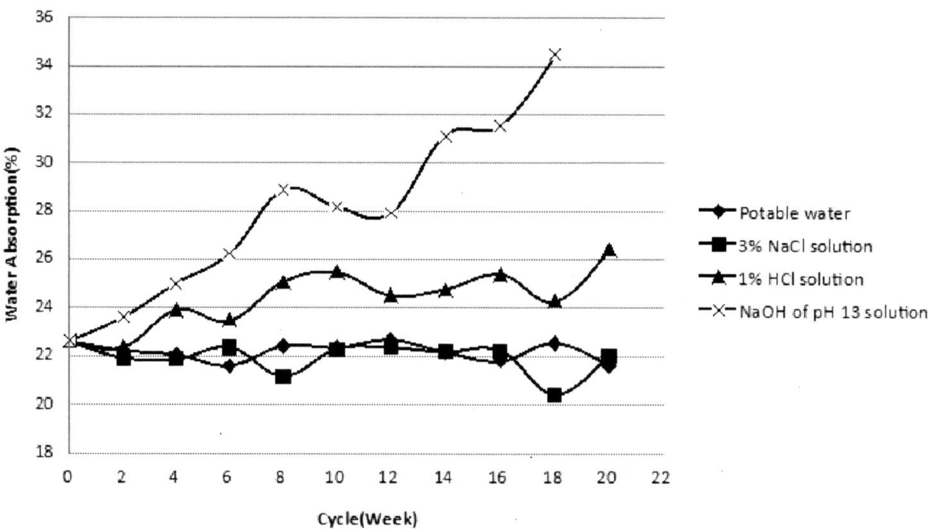

Figure 3.1. Water absorption of OPS under wetting and drying cycles in various solutions [39, with modification].

The water absorption of OPS aggregates after being subjected to various deleterious solutions under wetting and drying conditions is shown in Figure 3.1 [39]. OPS aggregates under 1% HCl and NaOH solutions have shown a wide increasing pattern in the water absorption values in comparison to potable water and 3% NaCl solution. Both acidic and alkaline solutions are expected to cause more internal pores to be formed with the increase in cycles of wetting and drying due to its damaging nature. The highest water absorption recorded was from the specimens exposed to the NaOH solution.

3.6. AIV of OPS Aggregate Treated in Water, HCl, NaCl and NaOH Solutions

The AIV is a primary test used to determine the toughness of an aggregate. It is measured as the resistance of the aggregate to failure by impact. The toughness property of aggregate is required when aggregate is used in pavement construction or subjected to high wearing. Toughness is defined as the resistance of aggregate to failure by impact. The impact value is a measure of toughness. The maximum permissible values of AIV for normal construction, wearing surface such as runways, roads and heavy duty floors are 45%, 30% and 25% respectively.

The AIV of OPS aggregates after being subjected to wetting and drying cycles in various solutions is presented in Figure 3.2 [39]. OPS aggregate has superior quality in comparison to other normal weight aggregates like granite, sandstone, limestone etc. in terms of AIV. The initial AIV of OPS aggregates is measured as 3.05% which indicates that the OPS aggregate has good energy absorbing capacity. Both potable water and 3% NaCl solution do not cause any damaging effect on OPS aggregates for AIV tests as seen in Figure 3.2. After exposure to water and 3% NaCl solution, the AIV ranged from 2.50% to 3.18% and 2.52 to 3.37% respectively.

On the other hand, when the OPS aggregates were exposed to 1% HCl solution, there was a large difference between the initial and final impact value. After 20 cycles of exposure, the highest AIV recorded was 7.48%, which is double the initial value. This indicates that the acid has a harmful effect on the OPS aggregates. This may be due to the fact that acidic solutions have a corrosive effect on the internal composition of the OPS aggregates. In previous research, it was reported that 1% HCl solution was found to be more corrosive compared to other types of acids such as 5% acetic acid, 1% lactic acid and 1% sulfuric acid on concrete [42]. Therefore its effect on raw natural material like OPS is expected to be more dominant. The cellulose membrane, which provides structural integrity and toughness for the OPS aggregate, is expected to be reduced by the acidic solution. Although the acidic solution caused the most damage, the AIV for OPS aggregates under acidic solution still falls within the range of 3.16 to 7.48%, which is much lower compared to that of conventional granite aggregate having an AIV of about 13.95% [1].

The OPS aggregate in alkaline solution (NaOH) recorded a decreasing pattern of AIV from the start of the wetting and drying cycle. From the previous section, it was found that the OPS aggregates under NaOH solution had an increasing water absorption value with increasing cycles and also recorded the highest water absorption value. The high amount of water in the aggregate pores could have caused the aggregate to become more flexible and therefore performed better as a shock absorber. This ultimately reduces the AIV of OPS aggregates.

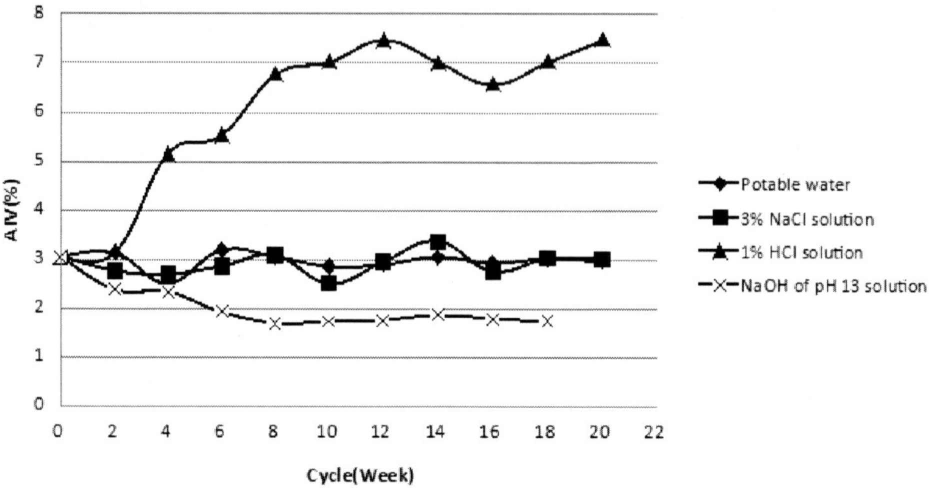

Figure 3.2. AIV of OPS aggregates under wetting and drying cycles in various solutions [39, with modification].

3.7. ACV of OPS Aggregate Treated in Water, HCl, NaCl and NaOH Solutions

The ACV test is carried out to study the crushing properties of aggregates in bulk, which indicate the resistance to pulverisation. For roads, the maximum crushing value of aggregate is to be permitted to 30%, and 45% for other structural applications. The conventional normal weight aggregates show crushing values from 12 to 24%.

The ACV of OPS aggregates after being exposed to cyclic wetting and drying in various solutions is presented in Figure 3.3 [39]. The tests were conducted at longer intervals as shown in this figure. The initial ACV value of OPS aggregate is 2.33%. Compared to initial ACV value of OPS aggregate, the wetting and drying cycles under potable water, salt solution and alkaline solution was found to not have any obvious damaging effect. The ACV of OPS under potable water as control condition ranges between 1.88 to 2.42%. When the aggregates were exposed to the salt and alkaline solution, the ACV ranged from 1.94 to 2.59% and 1.94 to 2.48% respectively.

However, when the OPS aggregates were exposed to wetting and drying cycles under acidic solution (1% HCl), the ACV recorded in the fourth cycle was more than double the initial value. In the 20th cycle, the ACV was found to be about 3.4 times higher compared to the initial value. A similar effect of aggressiveness by acidic solution on OPS aggregate was also shown in the previous section on the AIV results. The corrosive effect of the acid is believed to be the major cause. These results show that the internal cell and membrane structure of OPS undergo severe damage when exposed to acidic environment.

The ACV values of OPS aggregate without any treatment determined by other researchers are 4.67 to 10% [6, 26, 28].

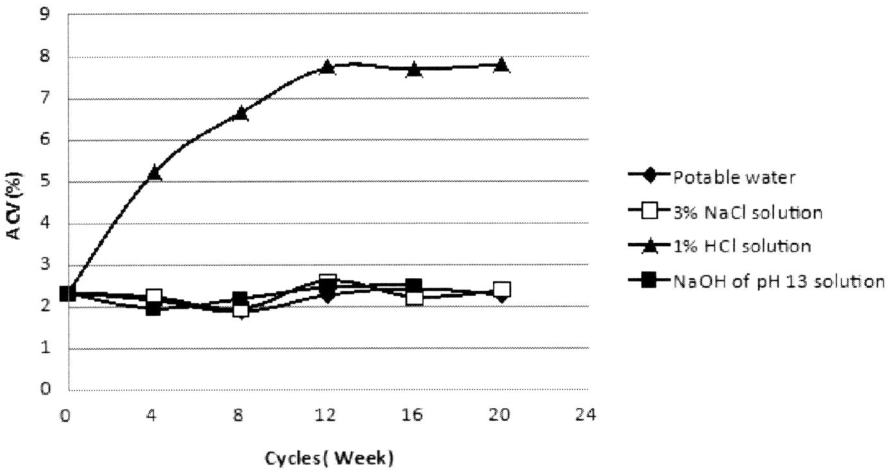

Figure 3.3. ACV of OPS under wetting and drying cycles in various solutions [39, with modification].

3.8. Chapter Remarks

An experimental investigation was conducted for the performance determination of oil palm shell (OPS) aggregates under aggressive conditions of wetting and drying for up to 20 cycles. Three aggressive solutions used were 1% HCl, 3% NaCl and NaOH of pH 13, and potable water as the control condition. The performance of OPS aggregates was evaluated based on the basic properties such as water absorption, aggregate impact value (AIV) and aggregate crushing value (ACV). The experimental investigation has revealed that OPS aggregate has performed well. Based on the results obtained from this research, the OPS aggregate was found to be fairly durable but also performed poorly under certain condition. Based on the obtained results, the following remarks were drawn:

1. The initial water absorption of OPS aggregates was found to be about 22.63%. When the aggregates were exposed to potable water and 3% NaCl there were no damaging effects observed. However, when the aggregates were exposed to 1% HCl and NaOH solutions, there was an increasing trend in water absorption values. The highest water absorption recorded was from the specimens exposed to the NaOH solution.
2. OPS aggregates have very low AIV and ACV, with initial values of 3.05% and 2.33% respectively. Results show that the OPS aggregates are durable under potable water, salt and alkaline conditions. It was only severely affected by the corrosive effect of the acidic solution.

References

[1] D.C.L. Teo. 2007. *A study on structural lightweight concrete using Oil Palm Shell (OPS) aggregate.* Ph.D. Thesis, Universiti Malaysia Sabah, Sabah, Malaysia.

[2] S. Mindess,J. F. Young and D. Darwin. 2003. *Concrete*. 2nd ed. USA: Prentice Hall.

[3] M.A. Mannan. 2001. *A study on the use of Oil Palm Shell in Lightweight Concrete*. PhD. Thesis. Universiti Malaysia Sabah, Sabah, Malaysia.

[4] C.H. Ng. 2010. *Studies on precast reinforced concrete floor panels using Oil Palm Shell aggregate*. PhD. Thesis. Universiti Malaysia Sabah, Sabah, Malaysia.

[5] D.C.L. Teo, M.A. Mannan, V.J. Kurian and C. Ganapathy. 2007. Lightweight concrete made from oil palm shell (OPS): structural bond and durability properties. *Building and Environment*. **42**. pp 2614-2621.

[6] D.C.L. Teo, M.A. Mannan and V.J. Kurian. 2006. Flexural behaviour of reinforced lightweight concrete beams with oil palm shell (OPS). *Journal of Advance Concrete Technology*, **4**(3), pp 459-468.

[7] D.C.L. Teo, M.A. Mannan and V.J. Kurian. 2006. Structural concrete using oil palm shell (OPS) as lightweight aggregate. *Turkish Journal of Engineering and Environmental Sciences*, **30**, pp 251-257.

[8] M.A. Mannan, J. Alexander, C. Ganapathy and D.C.L. Teo. 2006. Quality improvement of Oil Palm Shell (OPS) as coarse aggregate in lightweight concrete. *Building and Environment*. **41**(9). pp1239-1242.

[9] D.C.L. Teo, M.A. Mannan and V.J. Kurian. 2006. Influence of site curing on bond properties of reinforced lightweight concrete. *Journal of Civil Engineering Research and Practice*, **3**(1). pp 9-19.

[10] M.A. Mannan and C. Ganapathy. 2004. Concrete from an agricultural wastes-oil palm shell (OPS). *Building and Environment*. **39**(4). pp 441-448.

[11] M.A. Mannan and C. Ganapthy.2004. Destructive and non-destructive properties of OPS (Oil Palm Shell) lightweight concrete. *Journal of Civil Engineering Research and Practice*. **1**(2), pp55-66.

[12] M.A. Mannan and C. Ganapathy. 2002. Engineering properties of concrete with OPS as aggregate. *Construction and Building Materials*. **16**(1). pp 29-34.

[13] M.A. Mannan and C. Ganapathy. 2001. Mix design for oil palm shell concrete. *Cement and Concrete Research*. **31**(9). pp1323-1325.

[14] M.A. Mannan and C. Ganapathy. 2001. Long-term strengths of concrete with oil palm shell as coarse aggregate. *Cement and Concrete Research*. **31**(9). pp 1319-1321.

[15] M.M.U. Islam, K. H Mo, U.J Alengaram and M. Z Jumaat. 2015. Mechanical and fresh properties of sustainable oil plam shell lightweight concrete incorporating palm oil fuel ash. *Journal of Cleaner Production*, http://dx.doi.org/10.1016/j.jclepro.2015.12.051.

[16] U.J. Alengaram, B.A. Al-Muhit, and M.Z Jumaat. 2013. Utilization of oil palm kernel shell as lightweight aggregate in concrete - a review. *Construction and Building Materials*. **38**. pp 161-172.

[17] U.J. Alengaram, H. Mahmud, and M.Z. Jumaat. 2010. Comparison of mechanical and bond properties of oil palm kernel shell concrete with normal weight concrete. *International Journal of the Physical Sciences*. **5**(8). pp 1231-1239.

[18] U.J. Alengaram, H. Mahmud, and M.Z. Jumaat. 2011. Enhancement and prediction of modulus of elasticity of palm kernel shell concrete. *Materials & Design*. **32** (4). pp 2143-2148.

[19] K.Y. Foong, U.J. Alengaram, M.Z. Jumaat and K.H. Mo. 2015. Enhancement of the mechanical properties of lightweight oil palm shell concrete using rice husk ash and manufactured sand. *Journal of Zhejiang University SCIENCE A*. **16**(1). pp 59-69.

[20] K.H. Mo, U.J. Alengaram, M.Z. Jumaat, M.Y.J Liu and J. Lim. 2016. Assessing some durability properties of sustainable lightweight oil palm shell concrete incorporating slag and manufactured sand. *Journal of Cleaner Production.* **112**(1). pp 763-770.

[21] K.H. Mo, U.J. Alengaram, M.Z. Jumaat and S.P. Yap. 2015. Feasibility study of high volume slag as cement replacement for sustainable structural lightweight oil palm shell concrete. *Journal of Cleaner Production.* **91**. pp297-304.

[22] K.H. Mo, S.P. Yap, U.J. Alengaram, C.H. Bu and M.Z. Jumaat. 2014. Impact resistance of hybrid-fibre reinforced oil palm shell concrete. *Construction and Building Materials.* **50.** pp 499-507.

[23] P. Shafigh, U.J. Alengaram, H. Mahmud and M.Z. Jumaat. 2013. Engineering properties of oil palm shell lightweight concrete containing fly ash. *Materials & Design.* **49**. pp 613-621.

[24] P. Shafigh, M.Z. Jumaat and H. Mahmud. 2011. Oil palm shell as a lightweight aggregate for production high strength lightweight concrete. *Construction and Building Materials.* **25** (4). pp 1848-1853.

[25] S.A. Salam. 1982. Lightweight concrete made from palm oil shell aggregates and rice husk. *Proceedings of A Regional Seminar.* Universiti Putra Malaysia. Sept. 15-17. pp. 177-196.

[26] D.C. Okpala. 1990. Palm kernal shell as a lightweight aggregate in concrete. *Building and Environment.* **25**(4). pp 291-296.

[27] H.B. Basri, M.A. Mannan and M.F.M. Zain. 1999. Concrete using waste oil palm shells as aggregate. *Cement and Concrete Research.* **29**(4). pp 619-622.

[28] O.F. Okafor. 1988. Palm kernal shell as a lightweight aggregate for concrete. *Cement and Concrete Research.* **18**(6). pp 901-910.

[29] C.H Ng, Z Ideris, S.P Narayanan, M.A Mannan, V.J Kurian. 2007, Engineering properties of oil palm shell (OPS) hybrid concrete for lightweight precast floor slab, *Conventry University and UWM Center for By-products Utilization International Conference on Sustainable Construction Materials and Technologies*, UK, June 11-13.

[30] H. Mahmud, M.Z. Jumaat and U.J. Alengaram. 2009. Influence of sand/cement ratio on the mechanical properties of palm kernel shell concrete. *Journal of Applied Sciences,* **9**(9). pp 1764-1769.

[31] U.J Alengaram, H. Mahmud and M.Z. Jumaat. 2010. Development of lightweight concrete using industrial waste material, palm kernel shell as lightweight aggregate and its properties. *2nd International Conference on Chemical, Biological and Environmental Engineering (ICBEE 2010).* pp 277-281.

[32] P. Shafigh, M.Z Jumaat, H. Mahmud and N.A.A. Hamid. 2012. Lightweight concrete made from crushed oil palm shell: Tensile strength and effect of initial curing on compressive strength. *Construction and Building Materials.* **27**. pp 252-258.

[33] P. Shafigh, M.Z. Jumaat, H. Mahmud and U.J. Alengaram. 2013. Oil palm shell lightweight concrete containing high volume ground granulated blast furnace slag. *Construction and Building Materials.* **40**. pp 231-238.

[34] M.Y.J. Liu, C.P. Chua, U.J. Alengaram and M.Z. Jumaat. 2014. Utiliozation of palm oil fuel ash as binder in lightweight oil palm shell geopolymer concrete. *Advances in Materials Sciences and Engineering*, doi:10.1155/2014/610274.

[35] J. Wu, G. Liang, D. Hui, Q. Deng, X. Xiong, Q. Qiu, J. Liu, G. Chu, G. Zhou and D. Zhang. 2016. Prolonged acid rain facilitates soil organic carbon accumulation in a mature forest in Southern China. *Science of the Total Environment.* **544**. pp 94-102.

[36] Ng Chee Hiong, N. S. V Kameswara Rao and M. A. Mannan. 2011. Precast C-channel floor for rural and estate housings. *Malaysian Construction Research Journal* MCRJ. **9**(2).pp 63-72.

[37] B.H. Spratt. 1974. The structural use of lightweight aggregate concrete. *Cement Concrete Association.*

[38] FIP. 1983. *FIP Manual of Lightweight Aggregate Concrete.*, 2nd edition. London. Surry University Press.

[39] P. Ramasamy, H. Asrah, M.A. Mannan, D.C.L. Teo and K. Neglo, 2008. Performance of Oil Palm Shell (OPS) and OPS Concrete under Aggressive Conditions. Proceedings of EnCon2008, 2nd Engineering Conference on Sustainable Engineering, Infrastructures Development & Management, Kuching, Sarawak, Malaysia, December 18 -19.

[40] BS 812-112:1990 (Replaced By: BS EN 1097-2:1998). *Testing aggregates. Method for determination of aggregate impact value (AIV).*

[41] BS 812-110:1990 (Replaced By: *BS EN 1097-2:1998*). *Testing aggregates. Methods for determination of aggregate crushing value (ACV).*

[42] P. K. Mehta, 1985. Studies on chemical resistance of low water/cement ratio concretes. *Cement and Concrete Research.* **15**. pp 969-978.

Chapter 4

Mix Design of OPS Concrete

Keywords: Oil Palm Shell (OPS) aggregate, mix design, lightweight concrete, OPS hybrid concrete, OPS Geopolymer concrete

4.1. Introduction

The mix design of concrete is an art of selecting the optimum amount of ingredients to fulfil the required engineering properties as economically as possible. The mix design of lightweight concrete using oil palm shell (OPS) as coarse aggregate varies widely from conventional mix design as used for normal weight concrete (NWC) and lightweight concrete (LWC). In structural applications, the mix design of lightweight concrete is more complicated as it depends on the properties of the lightweight aggregate. The engineering properties of aggregate such as shape, surface texture, bulk density, absorption, hardness or inherent strength and grading are wide for different types of artificial (manufactured) and natural lightweight aggregates, and these are influencing factors in mix design. However, OPS aggregate being natural and organic solid waste has different shapes, sizes and thicknesses.

4.2. Basic Steps in Concrete Mix Design for Lightweight Concrete

Mix design is the art of selecting the optimum amount of ingredients in meeting the required engineering properties of concrete as economically as possible. The required engineering properties are workability on fresh concrete, strength and durability on hardened concrete.

In lightweight concrete, lightweight aggregates used are either naturally occurring or manufactured. Both types of lightweight aggregates have unique properties different from conventional normal weight aggregates, particularly on physical and mechanical characteristics. The type, grading and size of lightweight aggregates have an effect on the properties of lightweight concrete. Lightweight aggregates are generally weaker in strength than normal weight aggregates. Hence, a higher binder content is required to compensate.

Binder content used in lightweight concrete (LWC) in excess of 600 kg/m^3 has been reported [1, 2, 3]. According to [4], the binder (cement) content for LWC can be in the range of 250 to 560 kg/m^3 for achieving the compressive strength of 17.5 to 35.0 N/mm^2. The structural LWC has the density range from 1,440 to 2080 kg/m^3 [5]. The structural lightweight concrete is having a 28-day cylindrical compressive strength of 17.2N/mm^2 and the corresponding air-dry density of 1440-1850kg/m^3 as defined by ACI 213-87. In British Standard (BS3797-90), the air-dry density of LWC is not to exceed 2000kg/m^3. China Standard (JGJ 51-90) has mentioned that the oven-dry density of LWC is not to exceed 1950kg/m^3. CEB/FIP in Europe has mentioned LWC having the air-dry density of 1600-2000kg/m^3 and 28-day cube compressive strength from 20-70N/mm^2. In Italy (UNI 7548.1) standard, the density of LWC is not greater than 1850kg/m^3.

The maximum size of coarse aggregate used to produce structural LWC is generally limited to 14 mm [6]. Moreover, LWC also tends to have a higher amount of fine aggregate compared to NWC [7]. The workability is an essential requirement in the mix design of structural LWC, so LWC can be fully compacted at site [6]. Therefore, the optimum amount of water is crucial to produce good quality structural LWC.

Various mix design methods developed over the years are reported in literature. In arbitrary volume method is one such mix method where mix ingredients such as binder (cement), fine aggregate (sand) and coarse aggregate (crushed stone, or river pebble, or brick khoa made from crushing of burnt clay brick) are taken in volumes like 1: 2: 4 as an indication for grade 25 concrete (25 N/mm^2). The weight method in mix design is a relatively simple way to estimate the ingredient proportions based on the assumed density of concrete. In the absolute volume method, it is assumed that the volume of concrete is equal to the sum of absolute volume of all mix ingredients to produce 1m^3 concrete. The volume of entrapped air is deducted in this method. The absolute volume method as recommended [8] shows better accuracy over the weight method. Generally, the following equation is used to estimate the amount of all ingredients to produce 1.0 m^3 OPS concrete.

$$\frac{C}{1000\,\rho_C}+\frac{W}{1000\,\rho_W}+\frac{S}{1000\,\rho_S}+\frac{C.A}{1000\,\rho_{C.A}}=1 \qquad \text{(Eq. 4.1)}$$

Where,

C= OPC

W=water

S=sand (fine aggregate)

C. A= Coarse aggregate

ρ = suffix representing specific gravity of each ingredient

More ingredients can be added in this equation. This equation serves to start estimating the mix ingredients through trial mixes so as to arrive at the optimum mix of OPS concrete in fulfilling the required engineering properties.

The fundamental steps required on mix proportioning for both absolute volume and weight methods for a particular job are to evaluate [9]:

a) 28-day compressive strength
b) Slump as workability on fresh concrete

c) W/C ratio
d) Coarse aggregate including maximum size and quantity
e) Binder (cement) and type
f) Fine aggregate
g) Moisture correction for net free water content
h) Mixing, casting and testing as trial mixes
i) Determination of optimum mix through trail mixes with adjustments of above (step c to f)

The mix design of LWC varies widely from mix design of NWC as it depends upon the properties of coarse aggregate. The mix design of LWC used for structural purposes is more complicated because it depends on the type of lightweight aggregate used [10]. Several types of lightweight aggregates are in use. The mix design of LWC, especially when the new aggregate is used as coarse aggregate, has to be scrutinized very carefully. The aggregate shape, surface texture, bulk density, absorption, hardness or inherent strength and grading are wide for different types of lightweight aggregates, and they are also influencing factors in mix design. OPS aggregate is of different shapes, sizes and thicknesses.

4.3. Trial Step for Mix Design of OPS Concrete Using Conventional Mix Methods

In the preliminary investigation of OPS concrete mix, [11] used three procedures such as ACI mix design method as used for normal weight concrete and two other methods of mix design for lightweight concrete described by [6, 12].

4.3.1. Trial Step for Mix Design of OPS concrete according to the ACI method

The details of mix orders ACI-1, ACI-2 and ACI-3 based on the ACI mix design method, are given in Table 4.1. However, the 28-day compressive strengths were far lower compared with the target design strength of 28 N/mm^2 for OPS concrete. Though the superplasticizer, Sikament NN has been used in the mixes of ACI-2 and ACI-3, it has not increased the strength of OPS concrete.

Table 4.1. Mix Design for OPS concrete according to ACI Mix Design, superplasticiser of Sika NN type) [11, with modification]

Mix	Mix ingredients (kg/m^3) OPS of SSD			w/c ratio	SP (L/m^3)	Demould. density (kg/m^3)	28-day Comp. Strength (N/mm^2)
	OPC	Sand	OPS				Experiment
ACI-1	370	1010	315	0.53	0	1,870	13.65
ACI-2					3.5	-	6.30
ACI-3					1.5	1,884	13.15

Note: OPC-ordinary Portland cement.

4.3.2. Trial Step for Mix Design of OPS Concrete According to Short and Kenniburgh [6]

The details of mixes of OPS concrete based on the mix design method for lightweight concrete [6] are given in Table 4.2. The 28-day target design strength of 25N/mm^2 was not achieved in both mixes as shown in this table. Actually, the mix design procedure for lightweight concrete cited by [6] is for artificial lightweight aggregates using Leca, Foamed slag, Aglite and Lytag.

Table 4.2. Mix Design for OPS concrete according to the Method mentioned in [6], [11, with modification]

Mix	Mix ingredients (kg/m^3) OPS of SSD			w/c ratio	Slump (mm)	Demold. density (kg/m^3)	28-day Comp. Strength (N/mm^2)
	OPC	Sand	OPS				Experiment
Short et al.-1	555	710	305	0.60	Collap.	1,791	11.80
Short et al.-2	630	536	353			-	14.35

4.3.3. Trial Step for Mix Design of OPS Concrete According to Shetty [12]

A mix design method suggested [12] has been employed for the mixes of OPS concrete as shown in Table 4.3. The targeted design strengths at 28 days for mixes Shetty-1 and Shetty-2 are considered as 25N/mm^2 and 22N/mm^2 respectively. However, the results obtained are much less than the targeted design strength as shown in Table 4.3.

Table 4.3. Mix Design for OPS concrete according to the Method mentioned in [12], [11, with modification]

Mix	Mix ingredients (kg/m3) OPS of SSD			w/c ratio	Slump (mm)	Demould density (kg/m^3)	28-day Comp. Strength (N/mm^2)
	OPC	Sand	OPS				Experiment
Shetty-1	480	542	442	0.50	-	1,781	14.40
Shetty-2	400	564	460	0.60	Collap.	1,704	9.65

4.3.4. Trial Step for Mix Design of OPS Concrete According to the Absolute Volume Method

As discussed above, the ACI mix design method is mainly for normal weight concrete and the methods mentioned in [12] and [6] are for lightweight concrete using artificial lightweight aggregates. The OPS aggregate is a natural organic material and its properties are different from those of artificial lightweight aggregates such as Leca, foamed slag, Aglite and

Lytag in that the texture is smooth and the shapes are of different types. The targeted design strengths of OPS concrete have not been achieved by employing these established methods. Hence, it is concluded that established methods such as ACI for normal weight concrete and methods in [6, 12] for lightweight concrete are not applicable for the mix design of OPS concrete.

After failing to obtain the suitable mix design procedure for OPS concrete as discussed above, the trial steps have been performed using Equation 4.1 for the absolute volume method. As the above methods have failed to get the desired targeted design strengths, a large number of mixes (Table 4.4) have attempted to obtain the acceptable strength for OPS concrete. In these trial mixes, different variations are considered such as; mix proportions, cement content, w/c ratio, concrete density and stage of OPS aggregate (saturated surface dry/or, air dry condition). For lightweight concrete, the amount of cement content specified in the literature is in the range of 285 - 510kg/m^3 [13].

Table 4.4. Trial mix for OPS concrete using the absolute volume method [11, with modification]

Mix	Mix ingredients (kg/m^3)			w/c ratio	Stage of OPS	Slump (mm)	Demould density (kg/m^3)	28-day Comp. St. (N/mm^2)
	OPC	Sand	OPS					
A-1	325	553	553	0.65	SSD	0	1680	7.30
A-2	340	541	541	0.62	SSD	0	1655	9.90
A-3	465	577	460	0.55	Air dry	5	1780	11.85
A-4	465	512	493	0.55	Air dry	10	1755	11.50
A-5	480	576	470	0.54	SSD	-	-	10.40
A-6	500	620	395	0.55	Air dry	30	1825	15.70
A-7	520	645	447	0.38	SSD	5	1830	20.00
A-8	520	556	510	0.36	SSD	0	1815	18.75
A-9	545	447	430	0.55	Air dry	44	1725	14.75
A-10	555	688	278	0.60	Air dry	Collap.	1850	10.00
A-11	560	459	442	0.50	Air dry	30	1770	12.70
A-12	565	463	503	0.38	SSD	20	1780	18.60
A-13	585	725	293	0.50	Air dry	20	1870	18.80
A-14	600	492	438	0.41	SSD	25	1805	20.25
A-15	600	543	372	0.45	Air dry	40	1910	18.45
A-16	630	517	265	0.60	Air dry	Collap.	1790	11.50
A-17	675	554	284	0.50	Air dry	125	1845	19.10

4.3.5. Acceptable Mix Design for OPS Concrete

The acceptable mix proportions for OPS concrete with different ingredients are presented in Table 4.5. The air-dried sand has been used throughout the investigation for all types of mix designs and the required correction has been made to adjust the free water content.

Mix design of OPS concrete using FDN superplasticizer has shown better workability and acceptable compressive strength. OPS concrete can be made according to required workability and strength if FDN superplasticizer is available. OPS concrete with less cement content containing FDN superplasticizer (mix orders C-1 to C-14) has shown almost the same strength when compared with mixes B-1 to B-5 having a high cement content. The mix C-9 (Table 4.5) was used to build a footbridge and a thermally comfortable low cost house as elaborated in Chapter 8. The mix C-9 was also used to make precast SIConSofa blocks as discussed in Chapter 9.

Table 4.5. Acceptable mix proportions for OPS concrete [14]

Mix	Mix ingredients (kg/m^3) (OPS of SSD condition)					Concrete density (Kg/m^3)	Slump (mm)	Comp. st. (N/mm^2)
binder = 480kg/m^3, free w/c ratio = 0.41								
	OPC	Fly ash	$CaCl_2$	Sand	OPS	Demould.		28-day
B-1	480	0	0.00	821	370	1890-1905	<10	24.20
B-2	432	48	0.00					22.60
B-3	408	72	0.00					19.50
B-4	480	0	0.5%					23.45
B-5	480	0	1.0%					29.40
OPC=400kg/m^3, free w/c ratio=0.48, Superplasticiser of FDN type								
	OPC	SP (%)	Sand	OPS				
C-1	400	1	740	400		1859-1868	30	19.8
C-2		2					90	20.8
OPC=400kg/m^3, free w/c ratio=0.45								
C-3	400	1	756	408		1854-1851	<15	21.3
C-4		2					<20	21.0
OPC=420kg/m^3, free w/c ratio=0.48								
C-7	420	1	722	395		1873-1884	55	20.7
C-8		2					65	19.9
OPC=420kg/m^3, free w/c ratio=0.45								
C-9	420	1	739	399		1883-1894	45	21.2
C-10		2					50	19.8
C-11	420	1	748	403		1883-1890	25	20.9
C-12		2					45	20.7
OPC=400kg/m^3, free w/c ratio=0.46								
C-13	400	1	752	404		1882-1801	70	19.8
C-14		0					5	17.6

4.3.6. Workability of OPS Concrete

The slump, air content and compacting factor (C. F) as workability indicators of OPS concrete mixed with fly ash, without fly ash, with calcium chloride, and with or without FDN superplasticizer, are shown in Table 4.6. No attempt at air content determination has been performed for concrete containing FDN superplasticizer.

The workability of fresh concrete depends mainly on materials, mix proportion and environmental conditions. Generally, aggregates occupy about 60-80% of the total volume of concrete. Total specific areas of aggregate are minimised by the proper selection of size, shape and proportion of fine and coarse aggregates. The shape and surface texture of aggregates control the void content and affect the water requirement of concrete mix significantly. The fineness modulus of aggregate is a numerical index of fineness giving some idea of the mean size of that aggregate. The fineness modulus of aggregate is a prime indication for the most economic mix design to obtain the required strength and workability.

The workability of fresh concrete is influenced significantly by physical properties such as roughness, shape and texture of aggregates. The smoothness or roughness of aggregate reflects the surface texture. The glassy, smooth, granular, rough, crystalline, porous and honeycombed textures are the visual properties of the surface [15]. Overall, the slump values of OPS concrete are very low and this is due to the large amount of fine aggregate. The addition of fly ash and calcium chloride has not shown any significant difference in workability.

In fresh concrete, air content is relatively high in OPS concrete. OPS aggregates come in many shapes, such as roughly parabolic or semi-circular, flaky and other irregular shapes. OPS aggregates are also porous materials as observed by water absorption, and these may cause high air entrapment in fresh concrete. By the addition of fly ash and calcium chloride, OPS concrete has not shown any remarkable difference in air content compared with OPS concrete having no fly ash and calcium chloride. In the United Kingdom, the total air content of 4.5±1.5% by volume of a concrete mix is recommended [16].

OPS concrete with 1% FDN superplasticizer has shown better workability compared with the above mentioned mixes without this superplasticizer. The results of slump of OPS concrete (mix orders C-13, C-14) are based on cement content 400kg/m^3 and with w/c ratio of 0.46. The result of OPS concrete with 1% FDN, shows better slump and it is 14 times higher than OPS concrete with no FDN superplasticizer.

Table 4.6. Properties of fresh OPS concrete for different mixes [14]

Mix	Type of concrete	Properties of fresh concrete		
		Slump (mm)	C. F	Air (%)
OPS concrete				
B-1	OPS concrete with 0%FA	<10	0.81-0.83	4.9-5.1
B-2	OPS concrete with 10%FA		0.81-0.83	4.9-5.0
B-3	OPS concrete with 15%FA		0.82-0.84	4.8-4.9
B-4	OPS concrete with 0.5%$CaCl_2$		0.80-0.82	5.0-5.1
B-5	OPS concrete with 1%$CaCl_2$		0.80-0.83	4.8-4.9
C-13	OPS concrete with 1%FDN	70	0.83	N.A
C-14	OPS concrete with 0%FDN	5	0.78	N.A

4.3.7. Fresh and 28-Day Air-Dry Densities of OPS Concrete

Table 4.7 presents the fresh and 28-day air-dry densities of OPS concrete mixed with fly ash, without fly ash and with calcium chloride. Each result is the average of three samples.

Fresh concrete density is an early indication of the type of hardened concrete, whether it is ultra-lightweight or lightweight or structural lightweight or normal weight or heavy weight concrete. In this investigation, OPS concrete mixed both with and without fly ash, has produced a fresh density in the range of structural lightweight concrete. Normally, the fresh concrete density for lightweight concrete is around 100 to 200 kg/m^3 higher than the 28-day air-dry density [17]. An air-dry density is relevant for the weight of concrete in a structure. In this study, the 28-day air-dry density of OPS concrete is in the range of structural lightweight concrete. The density of structural lightweight concrete is within the range of 1,200 to 2,000 kg/m^3 [6, 13, 17, and 18].

The 28 day air-dry densities of OPS concrete containing 10% and 15% fly ash are 2% and 3% lower than OPS concrete having no fly ash, respectively. The low air-dry densities revealed in this study are due to the admission of fly ash with low specific gravity. Higher contents of fly ash reduce the concrete density.

The fresh and 28-day air-dry densities have not been influenced by the addition of calcium chloride of different content. Calcium chloride plays a vital role in the early strength gain of concrete only.

Table 4.7. Fresh concrete and 28-day air-dry densities of OPS concrete [14]

Mix	Type of concrete	Fresh and 28-day air-dry densities (kg/m^3)	
		Fresh density	28-day air-dry density
B-1	OPS concrete with 0%Fly ash	1958	1856
B-2	OPS concrete with 10%Fly ash	1923	1817
B-3	OPS concrete with 15%Fly ash	1910	1801
B-4	OPS concrete with 0.5%$CaCl_2$	1956	1850
B-5	OPS concrete with 1%$CaCl_2$	1955	1848

4.4. Mix Design of OPS Hybrid Concrete for Precast Applications

A number of researchers in recent years have further investigated the applicability of OPS in LWC. The attempts at making use of and improving the developed concrete made with OPS as structural LWC should not stop until and unless it is proven otherwise.

Hence, [19] has developed a mix design using both OPS and conventional crushed granite as coarse aggregate for structural LWC. The resulting concrete made with both OPS and granite as coarse aggregate is known as OPS hybrid concrete. He has used OPS of two types of shell thickness ranging less than 3mm and more than 3mm. His mix design of OPS hybrid concrete is based on the focus of 1-day and 28-day compressive strength, and a modulus of elasticity at 28-day of more than 10N/mm^2, 20N/mm^2 and 15GPa respectively, using OPS of less than 3mm as the majority shell thickness. Similarly, 15N/mm^2, 35N/mm^2

and 20GPa are using OPS of more than 3mm as the majority shell thickness. It is noted that [20] their investigation on OPS concrete has developed the highest modulus of elasticity of 10.9GPa. [19] have conducted several trial mixes using OPS and granite as coarse aggregates by varying the cement, OPS and sand content, water-cement ratio and the amount of superplasticizer. The acceptable trial mixes are then identified and finally, the workability, strength and density are taken into consideration when adopting the optimum OPS hybrid concrete mixes. For comparison purposes, OPS concrete (a LWC) and normal weight concrete (a NWC) are produced based on the same w/c ratio. Comparison among OPS hybrid concrete (LWC-1 and LWC-2), OPS concrete (LWC-3) and normal weight concrete (NWC) are conducted on the fresh and hardened properties.

4.4.1. Materials

[19] have used ordinary Portland cement (ASTM Type I) as the binder, river sand as the fine aggregate and, OPS and granite as the coarse aggregates. The OPS have been obtained in two batches and sieved as coarse aggregate of passing 12.5 mm and retained on a 4.75 mm sieve. It is found that one batch has a major shell thickness of 2.5 to 2.9 mm (known as shell thickness with majority of less than 3mm) whereas the other 3.1 to 3.5 mm (known as shell thickness with majority of more than 3mm). The granite aggregate of 19 mm maximum size has been used. The properties of OPS and granite are shown in Table 4.8. Type-F naphthalene sulphonate formaldehyde condensed based superplasticiser (SP) in aqueous form conforming to [21] has been used to increase the workability.

Table 4.8. Properties of OPS and crushed granite [19]

Descriptions	Coarse Aggregate		
	OPS		Crushed Granite
Physical Properties			
Maximum size, mm	12.00		19.00
Largest quantity of shell thickness, mm	2.5-2.9	3.1-3.5	-
Specific gravity (SSD)	1.33	1.34	2.63
Water absorption capacity (24 hrs), %	22.63	21.30	2.04
Moisture content as received, %	-	11.68	0.78
Bulk density (compacted), kg/m^3	599	646	1,482
Fineness modulus, F. M.	6.23	6.30	6.94
Mechanical Properties			
Los Angeles Abrasion Value (AAV), %	-	2.90	11.85
Aggregate Impact Value (AIV), %	3.05	6.60	12.60
Aggregate Crushing Value (ACV), %	2.33	7.12	15.95

4.4.2. Mix Design Development for OPS Hybrid Concrete

[19] have used the absolute volume method recommended by [8] for the mix design of OPS hybrid concrete using the following equation. It is used to estimate the quantities of all ingredients to produce 1.0 m^3 of concrete as indicated in the following equation:

$$\frac{C}{1000\,\rho_C}+\frac{W}{1000\,\rho_W}+\frac{S}{1000\,\rho_S}+\frac{OPS}{1000\,\rho_{OPS}}+\frac{G}{1000\,\rho_G}=1 \qquad \text{(Eq. 4.2)}$$

Where,

C= OPC
W=water
S=sand (fine aggregate)
OPS = oil palm shell as coarse aggregate
G= granite as coarse aggregate
ρ = suffix representing specific gravity of each ingredient

Various trial mixes are shown in Table 4.9 where the acceptable mixes has been identified. Finally, an optimum mix has been selected for further investigations.

The slump values in the range of 20-100 mm are suitable for structural beams, columns and floors [22]. Nevertheless, the rate of slump loss is very much affected by the rate of absorption of mix water by the lightweight aggregate and ambient condition. The poor workability of fresh OPS concrete without the use of superplasticizer at a given water/cement ratio is documented [11].

To start the trial mix, a mix designed for OPS concrete by [23] has been first adopted to verify the properties of OPS. In mix [23], the cement content used, 510 kg/m^3 is considered high even though maximum cement content for LWC can reach 560 kg/m^3 to achieve compressive strength of 35 MPa [24].

Table 4.9. Trial mixes of OPS hybrid concrete [19, with modification]

Description	Properties
Range of materials used in 1m^3 concrete	
OPC (binder), kg/m^3	380-510
River sand as fine aggregate, kg/m^3	620-848
OPS as coarse aggregate, kg/m^3	226-406
Granite as coarse aggregate (10, 20, 30, 40 and 50% replacement order of OPS), kg/m^3	99-492
w/c ratio	0.38-0.45
Superplasticiser, litres/100kg cement	1.2 – 2.0
Range of fresh properties	
Slump, mm	30-135
Fresh density, kg/m^3	1888-2180
Range of hardened densities	
Demoulded density, kg/m^3	1868-2161

Description	Properties
28-day air dry density, kg/m^3	1805-2093
Range of compressive strength	
3-day compressive strength, N/mm^2	14.11-25.80
28-day compressive strength, N/mm^2	25.04-38.60

The typical compressive strength of OPS concrete in past investigations without the inclusion of supplementary cementing materials (SCM) and fibers was found in the range of 25 to 30 MPa. The modulus of elasticity, E value was as low as 8GPa [25]. The E value should be increased to improve the stiffness of structural member. OPS alone cannot increase the E value of concrete, therefore conventional crushed granite aggregate was used as partial coarse aggregate, keeping the density of OPS hybrid concrete as lightweight concrete. Therefore, crushed granite aggregate was used with 10, 20, 30, 40 and 50% replacement of OPS aggregate in OPS hybrid concrete [19]. Table 4.9 shows the trial mix details and obtained engineering properties of this hybrid concrete.

4.4.3. Acceptable Mix Design for OPS Hybrid Concrete

[19] have obtained an acceptable mix design with further modification of trial mixes. The acceptable mixes are shown in Tables 4.10 and 4.11 for different thicknesses of OPS shell. Trial mixes with a slump value of not more than 100 mm as per [26] and compressive strength of more than 35 MPa were considered as acceptable mixes for LWC-1.

The cement content for such acceptable mixes is 430 and 450 kg/m^3, and this is within the permissible range for LWC. The resulting 28-day air-dry densities below 2,000 kg/m^3 indicate that OPS hybrid concrete can be classified as LWC. This is comparable to other LWC such as those made from expanded clay, expanded polystyrene, and crushed lightweight bricks [2, 27, 28]. The workability and strength for different applications of LWC are taken into consideration when selecting the optimum mix for OPS hybrid concrete.

The concrete mix made using an OPS of 3.1 – 3.5 mm thick shell is known as LWC-1. This mix proportion has a cement content of 450 kg/m^3 which is in the usual range of LWC [24]. Also, the permissible partial replacement of OPS aggregate by granite is found to be 30% in order to maintain concrete density below 2,000 kg/m^3. The superplasticizer of 1.7 litres per 100 kg of cement is within the recommended dosage provided by the manufacturer. The slump value, 90 mm is also satisfactory for structural LWC as per [29]. The fine and coarse aggregates occupy approximately 74% of the total volume of OPS hybrid concrete.

Insights of OPS hybrid concretes known as LWC-1 and LWC-2 with major shell thicknesses 3.1 – 3.5 mm and 2.5 – 2.9 mm, OPS concrete (LWC-3) with major shell thickness of 3.1 -3.5 and normal weight concrete (NWC) are shown in Figures 4.1 to 4.4 respectively. The volumetric shape of different types of concretes is depicted in Figure 4.5. Approximately 73% and 72% of OPS is used as coarse aggregate by volume in LWC-1 and LWC-2 respectively. A range of 50 – 100% of lightweight coarse aggregate as replacement of total coarse aggregate (by volume) has been used in concrete production [30]. OPS with 2.5 – 2.9 mm shell thicknesses have caused the significant reduction in the compacted bulk density.

The mix designs developed by [19] for LWC-1, LWC-2 and LWC-3 have been used for detailed investigation of prototype precast floor panels of 3 to 8m long as discussed in Chapter 7.

Table 4.10. Acceptable mix proportions for OPS hybrid concrete (LWC-1) using 3.1 - 3.5 mm thick OPS [19, with modification]

Description	Properties
Range of materials used in 1m^3 concrete	
OPC (binder), kg/m^3	430-450
River sand as fine aggregate, kg/m^3	620-637
OPS as coarse aggregate, kg/m^3	337-346
Granite as coarse aggregate, kg/m^3	309-317
w/c ratio	0.38-0.40
Superplasticiser, litres/100kg cement	1.2-1.7
Range of fresh properties	
Slump, mm	55-100
Fresh density, kg/m^3	2002-2085
Range of hardened densities	
Demoulded density, kg/m^3	-
28-day air dry density, kg/m^3	1917-1998
Range of compressive strength	
3-day compressive strength, N/mm^2	25.63-25.80
28-day compressive strength, N/mm^2	35.80-38.53

Table 4.11. Acceptable mix proportions for OPS hybrid concrete (LWC-2) using OPS of 2.5 – 2.9 mm shell thickness [19, with modification]

Description	Properties
Range of materials used in 1m^3 concrete	
OPC (binder), kg/m^3	450
River sand as fine aggregate, kg/m^3	629
OPS as coarse aggregate, kg/m^3	320
Granite as coarse aggregate, kg/m^3	269
w/c ratio	0.38
Superplasticiser, litres/100kg cement	1.7
Range of fresh properties	
Slump, mm	95
Fresh density, kg/m^3	1997
Range of hardened densities	
Demoulded density, kg/m^3	-
28-day air dry density, kg/m^3	1912
Range of compressive strength	
3-day compressive strength, N/mm^2	16.56
28-day compressive strength, N/mm^2	22.97

4.4.4. Properties of Fresh Concrete

The fresh concrete properties for all types of concrete are studied in [19]. The fresh properties included are the fresh density and slump as shown in Table 4.12. The workability of a mix is influenced by the physical properties of aggregate such as its roughness, surface texture and shape. The slump of LWC-1, LWC-2, LWC-3 and NWC is shown in Table 4.12. At 0 min, the slump obtained for OPS hybrid concrete (LWC-1 and LWC-2) is in the range of 90 – 95 mm, which indicates that OPS hybrid concrete has good workability as proposed by [31]. The intrusion of crushed granite aggregates as a partial replacement to OPS aggregate reduces water demand as OPS possesses larger surface areas. Teo (2007) [23] has recommended that mixing concrete made with OPS should be done under shade and protected from direct sunlight to prevent slump loss. OPS concrete should be protected from exposure to direct sunlight during transportation and casting should be completed within 30 minutes after mixing [23]. These recommendations favor the real production practice for OPS hybrid concrete used in the manufacture of precast elements under good quality control at the precast plants.

LWC-3 exhibits medium workability which ranges from 55 to 75 mm. The shape of OPS which is flaky and elongated, leads to possible interlocking of particles which could greatly reduce the workability. However, NWC is found to have a much higher workability, having slump values with ranging from 120 – 140 mm. The use of larger-sized crushed granite aggregates compared to OPS aggregate, reduces water demand. With larger-size granite aggregate, the surface area for hydration becomes smaller which allows higher workability.

The density or unit weight of concrete indicates whether it falls under normal weight or lightweight concrete. The fresh density of LWC-1, LWC-2 and LWC-3 are in the range of 1880-2020 kg/m^3 (Table 4.12). [32] has found that the fresh density of LWC can reach the maximum value at 2,200 kg/m^3. Also, this fresh density was about 70 to 100 kg/m^3 higher than its 28-day air-dry density. As expected, the fresh density of LWC-1 and LWC-2 is approximately 6% higher than that of LWC-3 due to the 30% replacement with granite aggregate which has higher density compared to that of OPS aggregate. The fresh density of NWC is about 18% higher compared to LWC-1 and LWC-2.

Table 4.12. Comparison of fresh properties of OPS hybrid concrete with other concrete types [19]

Description	Concrete symbol	Fresh properties	
		Slump (mm)	Density (kg/m^3)
OPS hybrid concrete using OPS of 3.1-3.5mm shell thickness	LWC-1	90-95	1990-2020
OPS hybrid concrete using OPS of 2.5-2.9mm shell thickness	LWC-2	90-95	1984-2004
OPS concrete using OPS of 3.1-3.5mm shell thickness	LWC-3	55-75	1880-1905
Normal weight concrete	NWC	120-140	2340-2380

4.4.5. Properties of Hardened Concrete

The hardened concrete properties for each type of concrete are shown in Table 4.13 [19]. Among them are the 28-day air-dry density, 1-day and 28-day compressive strength, and 28-day modulus of elasticity. Both OPS hybrid concrete (LWC-1 and LWC-2) and OPS concrete have met the requirements as structural lightweight concrete in terms of air-dry density, compressive strength and modulus of elasticity. The significant improvement on the modulus of elasticity is seen with LWC-1. The properties of NWC concrete is shown for comparison purposes. Granite is used as an identical volume replacement with OPS aggregate for this NWC.

Table 4.14 shows the compressive strength development of LWC-1, LWC-2, LWC-3 and NWC with initial curing in water up to 365 days. For LWC-1, the compressive strength is found to be approximately 38% higher compared to that of LWC-3 at 365 days. In general, the strength development of concrete is very much influenced by the inter-particle bond, porosity and strength of the paste, and strength of the aggregates [33].

Table 4.13. Hardened properties of LWC-1, LWC-2, LWC-3 and NWC [19]

Description	Concrete symbol	Air-dry density at 28-day (kg/m³)	Comp. st. (N/mm²)		Modulus of elasticity, E at 28-day (GPa)
			1-day	28-day	
OPS hybrid concrete using OPS of 3.1-3.5mm shell thickness	LWC-1	1917	17.30	38.53	21.10
OPS hybrid concrete using OPS of 2.5-2.9mm shell thickness	LWC-2	1912	11.31	22.97	16.28
OPS concrete using OPS of 3.1-3.5mm shell thickness	LWC-3	1805	13.77	27.90	16.37
Normal weight concrete	NWC	2290	41.18	58.33	32.58

Table 4.14. Development of compressive strength with initial period of curing in water [19]

Type of concrete	Development of compressive strength (N/mm²)				
	1-day	3-day	7-day	28-day	365-day
LWC-1	17.30	25.80	35.97	38.53	39.50
LWC-2	11.31	16.56	21.02	22.97	24.10
LWC-3	13.77	22.10	24.35	27.90	28.60
NWC	41.18	43.56	48.50	58.33	58.90

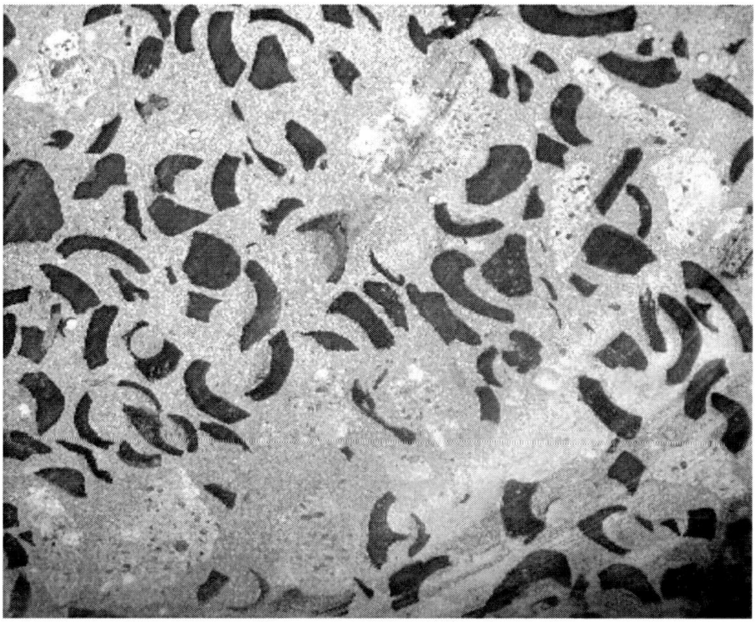

Figure 4.1. Insight of OPS hybrid concrete (LWC-1) using OPS of 3.1-3.5mm of majority shell thickness [34, with permission].

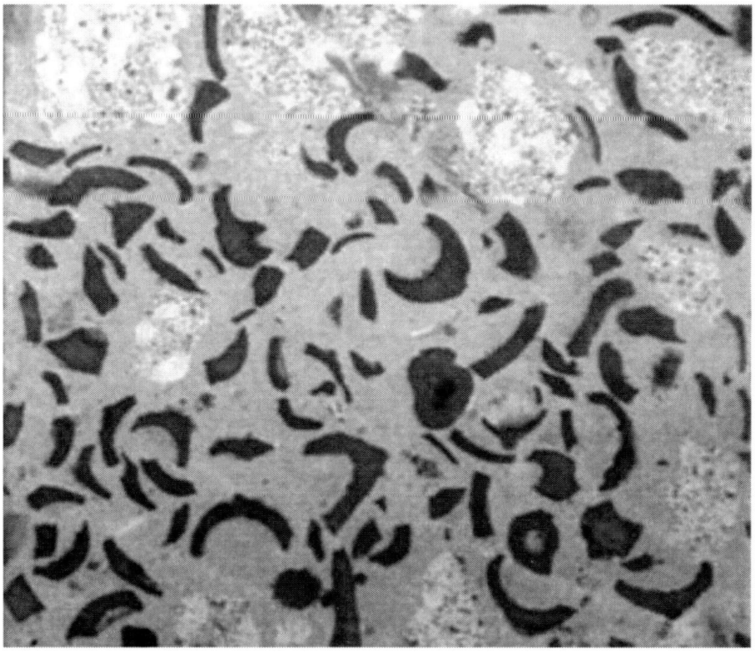

Figure 4.2. Insight of OPS hybrid concrete (LWC-2) using OPS of 2.5-2.9mm of majority shell thickness [19].

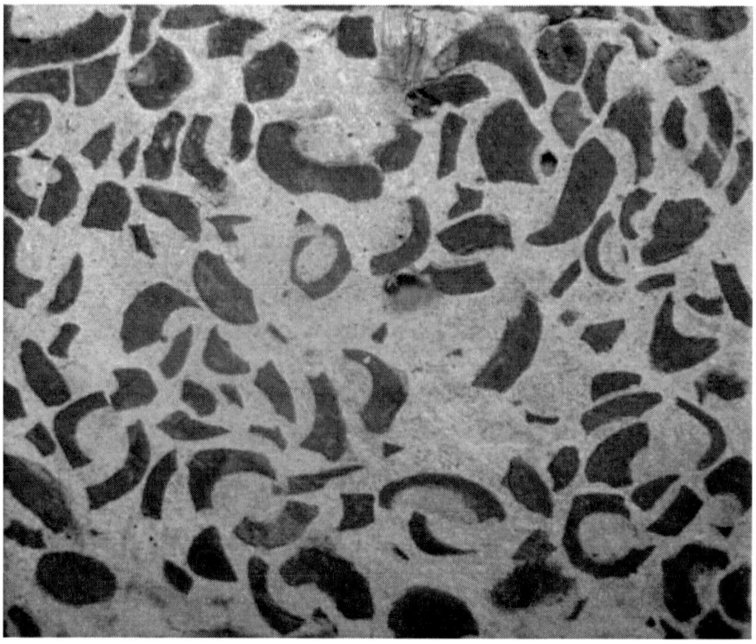

Figure 4.3. Insight of LWC-3 using OPS of 3.1-3.5mm of majority shell thickness [34, with permission].

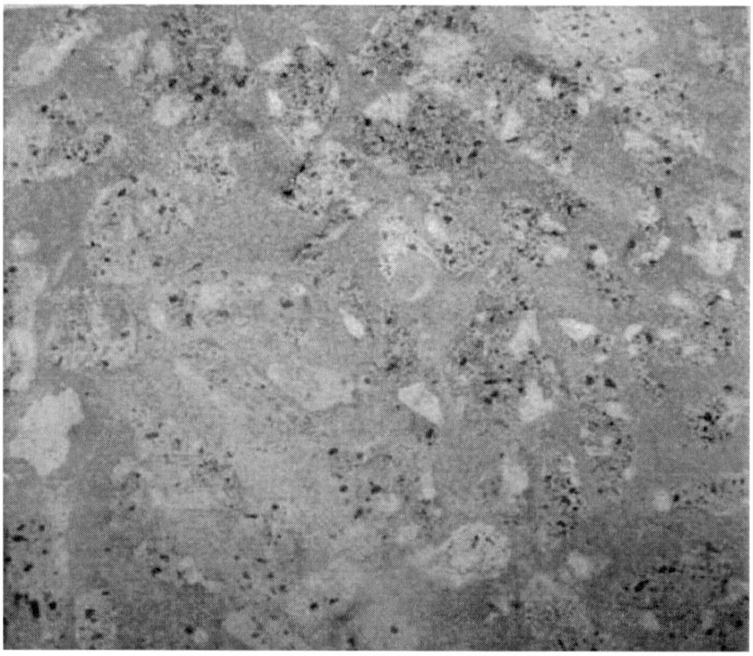

Figure 4.4. Insight of NWC concrete using crushed granite aggregate [34, with permission].

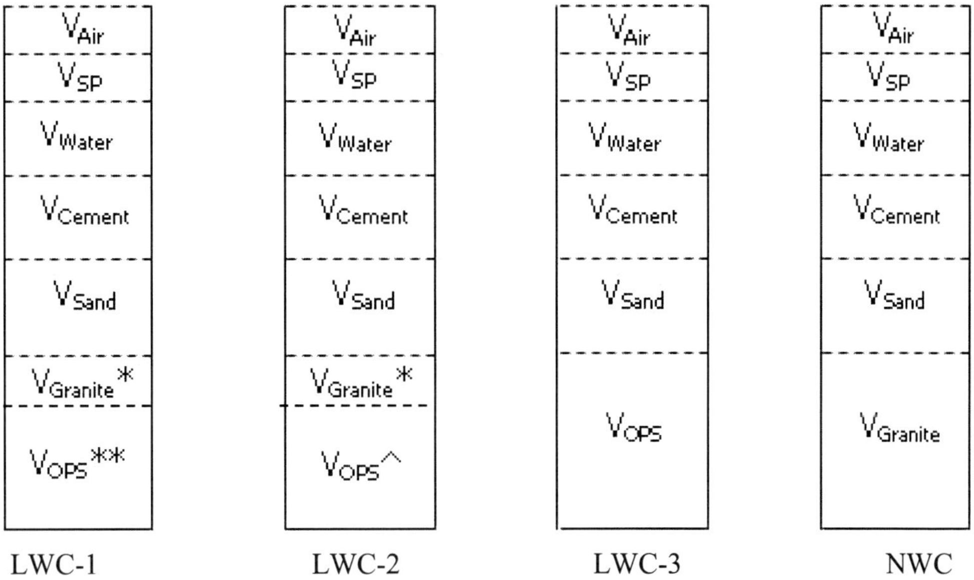

V_{OPS}** = 70% OPS with 3.1 – 3.5 mm shell thickness

$V_{Granite}$* = 30% granite

V_{OPS}^ = 70% OPS with 2.5 – 2.9 mm shell thickness

Figure 4.5. Typical volume of different types of concretes [19].

4.5. Mix Design of OPS Geopolymer Concrete

[35] have developed a mix design for OPS geopolymer concrete as lightweight concrete. They have used rice husk ash (RHA), metakaolin (MK) and slag (GGBS) as supplementary cementing materials (SCM) with an alkaline solution. The specific gravities of RHA, MK, and GGBS are 2.30, 2.89 and 2.50 respectively. The corresponding specific surface areas are 2981.2, 405 and 4315.8 m^2/kg. Fifty percent palm oil clinker (POC) sand and fifty percent manufactured sand are used as fine aggregate. The maximum amount of OPS as coarse aggregate used is 278.05 kg and replacement of OPS is done with granite at 0, 20, 40 and 80% replacement. They have used an alkaline activator using 14M (molarity) of NaOH solution. This solution is made 24 hours before casting concrete. This solution is mixed with sodium silicate at a ratio of 1:2.5 by weight before casting. The alkaline activator to binder ratio at 0.5 and water to binder ratio at 0.25 are maintained. According to [36], water to geopolymer solid ratio by mass is in the range from 0.16 to 0.24 and can produce very stiff to high workability concrete. However, the oven dry density revealed in this study is below 2000kg/m^3. The 3-day, 7-day and 28-day compressive strengths obtained depending on mixes are in the range of 24- 35N/mm^2, 24.3-35.2N/mm^2 and 25 -36N/mm^2 respectively. OPS geopolymer concrete having 100% OPS as coarse aggregate has shown compressive strength

of 24, 24.3 and 25N/mm^2 at 3-day, 7-day and 28-day respectively. This shows that the alkaline activator is superior in terms of strength gain over a short period.

4.6. Mix Design of OPS Concrete Using Different Supplementary Cementitious Materials and Fibers

[37] have investigated OPS concrete mixed with fly ash and granite as a partial replacement of cement and OPS respectively. They have used 550kg/m^3 cement, 900kg/m^3 sand and 504kg/m^3 coarse aggregate and 5.5 kg/m^3 SP. They have replaced cement with fly ash at 10, 30 and 50%. The 19% and 35% weight of coarse aggregate is replaced with granite. Their w/c ratio is 0.29. The developed slump values are in the range of 140-240mm, with air dry densities of 1886-2011kg/m^3. The 28-day compressive strengths are in the range of 30.04 to 44.49N/mm^2 depending upon fly ash inclusion. Higher fly ash content has reduced the compressive strength of OPS concrete as compared with OPS concrete having no fly ash.

[38] have used palm oil fuel ash (POFA) as supplementary cement material at 0, 10, 30, 50 and 70% OPC replacement in their mix design on OPS concrete. They have noticed that a higher percentage of POFA has shown an agglomeration effect on fresh concrete. The early strength gain is lower in OPS concrete having POFA than control OPS concrete having no POFA. But higher strength gain is noticed at later ages with POFA mixed OPS concrete.

[39] have used 20 and 60% slag (GGBS) as cement replacement and 0.3 and 0.6% steel fibers as volume fraction in the mix design for OPS concrete. In addition, they have used 560kg/m^3 cement, 1010kg/m^3 manufactured sand, 350kg/m^3 OPS, 168kg/m^3 water and 2.52kg SP. They have developed a 28-day compressive strength of about 32 to 41N/mm^2 depending on ingredients and other factors. The 20% slag and 0.6% steel fibers in OPS concrete mix with a water curing condition has shown the highest compressive strength.

4.7. Chapter Remarks

This chapter describes how mix design is developed for structural applications. The mix designs for OPS concrete are made in several ways such as using basic materials, hybrid coarse aggregate, alkaline activator as Geopolymer concrete, supplementary cement materials and fibers. The 28-day highest compressive strength and modulus of elasticity of this OPS concrete are 44.49N/mm^2 and 21.10GPa respectively. It shows that OPS concrete is comparable to conventional concrete as a structural concrete.

References

[1] J. A. Rossignolo, M. V. C. Agnesini. 2004. Durability of Polymer-modified Lightweight Aggregate Concrete. *Cement and Concrete Composites*. **26**(4). pp375 – 380.

[2] K. G. Babu, S. D. Babu. 2003. Behaviour of Lightweight Expanded Polystyrene Concrete Containing Silica Fume. *Cement and Concrete Research*. **33**(5). pp755 – 762.

[3] K. Dhir, R. G. C. Mays, H.C. Chua. 1984. Lightweight Structural Concrete with Aglite Aggregate: Mix Design and Properties. *The International Journal of Cement Composites and Lightweight Concrete*. **6**(4). pp249 – 261.
[4] ACI 613. 1959. Recommended Practice of Selecting Proportions for Structural Lightweight Concrete (ACI 613-59). *ACI Manual of Concrete Practice*. Detroit: American Concrete Institute.
[5] Martha G. Van Geem. 1990. Heat transfer characteristics of a recently developed lightweight structural concrete, *Insulation Materials, Testing, and Applications, ASTM STP 1030*, D. L. McElory and J. F. Kimpflen, Eds. *American Society for Testing and Materials*, Philadelphia. pp 437-463.
[6] A. Short, W. Kinniburgh. 1978. *Lightweight Concrete*, 3rd edition. London. Applied Science Publishers Ltd.
[7] S. Mindess, J. F. Young, D. Darwin. 2003. *Concrete*. (2nd edition). New Jersey: Prentice Hall.
[8] M. L. Ghambir. 1995. *Concrete Technology*. (2nd edition). New Delhi: Tata Mc-Graw Hill Publishing Company Limited.
[9] Michael S. Mamlouk, John P. Zaniewski. 1999. *Materials for Civil and Construction Engineers*, An imprint of Addision Wesley Longman, Inc.
[10] D. F. Orchard. 1979. *Concrete Technology*, all volumes, 4th edition. London. Applied Science Publishers Ltd.
[11] M. A. Mannan, C. Ganapathy. 2001. Mix Design for Oil Palm Shell Concrete. Cement and Concrete Research. **31**(9). pp1323 -1325.
[12] M. S. Shetty. 1993. *Concrete Technology*. India. S. Chand and Company ltd.
[13] M. Sidney, J. F. Young. 1981. *Concrete*. Englewood Cliffs, New Jersey. Prentice-Hall Inc.
[14] M. A. Mannan. 2001. *A study on the use of Oil Palm Shell in Lightweight Concrete*. PhD. Thesis. Universiti Malaysia Sabah.
[15] M. L Gambhir. 2004. *Concrete Technology*, 3rd edition. India. Tata McGraw-Hill Publishing Company limited.
[16] R. J. Salter. 1979. *Highway Design and Construction*. London. The Macmillan Press Ltd.
[17] Clarke John L. 1993. *Structural Lightweight Aggregate Concrete*. Blackie Academic & Professional- an Imprint of Chapman & Hall.
[18] B. H. Spratt. 1974. *The structural use of lightweight aggregate concrete*. Cement and Concrete Association.
[19] C. H. Ng. 2010, *Studies on precast reinforced concrete floor panels using Oil Palm Shell aggregate*. PhD. Thesis. Universiti Malaysia Sabah.
[20] H. Mahmud, M. Z Jumaat, U. J. Alengaram. 2009. Influence of Sand/Cement ratio on mechanical properties of palm kernel shell concrete. *Journal of Applied Sciences*. **9**. pp. 1764-1769.
[21] ASTM C 494. Standard Specification for Chemical Admixtures for Concrete. *Annual Book of ASTM Standards*.
[22] ACI 211. Standard Practice for Selecting Proportions for Structural Lightweight Concrete. Part 1. *ACI Manual of Concrete Practice*. Detroit: American Concrete Institute.

[23] D. C. L. Teo. 2007. *A study on structural lightweight concrete using Oil Palm Shell (OPS) aggregate*. Ph.D. Thesis. Universiti Malaysia Sabah.

[24] ACI 613. 1959. Recommended Practice of Selecting Proportions for Structural Lightweight Concrete (ACI 613-59). *ACI Manual of Concrete Practice*. Detroit: American Concrete Institute.

[25] M. A. Mannan, C. Ganapathy. 2002. Engineering properties of concrete with oil palm shell as coarse aggregate. *Construction and Building Materials*. **16**. pp 29-34.

[26] ASTM C 94, Standard Specification for Ready-Mixed Concrete. *Annual Book of ASTM Standards*.

[27] T. Y. Lo, H. Z. Cui, Z.G. Li. 2004. Influence of Aggregate Pre-wetting and Fly Ash on Mechanical Properties of Lightweight Concrete. *Waste Management*. **24**(4). pp333 – 338.

[28] J. Alduaji, K. Alshaleh, M. N. Haque, K. Ellaithy. 1999. Lightweight Concrete in Hot Coastal Areas. *Cement and Concrete Composites*. **21**(5). pp453 – 458.

[29] ASTM, C 143, Standard test method for slump of hydraulic-cement, *Annual Book of ASTM Standards*.

[30] K. M. A. Hossain. 2006. Blended Cement and Lightweight Concrete Using Scoria: Mix Design, Strength, Durability and Heat Insulation Characteristics. *International Journal of Physical Sciences*. **1**(1). pp5 – 16.

[31] P. Bartos. 1992. *Fresh Concrete*. Amsterdam: Elsevier Science Publishers.

[32] BS EN 12390-7: 2004. *Testing Hardened Concrete. Density of Hardened Concrete*. London: British Standards Institution.

[33] S. Popovics. 1998. *Strength and Related Properties of Concrete-a quantitative approach*. Canada. John Wiley & Sons.

[34] Ng Chee Hiong, N. S. V Kameswara Rao and M. A. Mannan. 2011. Precast C-channel floor for rural and estate housings. *Malaysian Construction Research Journal* MCRJ. **9**(2). pp 63-72.

[35] A. Sharmin, U. J. Alengaram, M. Z. Jumaat, S. M. A. Kabir, I. I. Bashar. 2015. Engineering properties of oil palm shell and palm oil clinker based geopolymer concrete. *Revista Técnica de la Facultad de Ingeniería Universidad del Zulia (REV TEC FAC ING UNIV)*. **38**. pp 34 – 48.

[36] B. V. Rangan. 2008. Low calcium fly ash based geopolymer concrete. Chapter 26 in *Concrete construction engineering handbook*. Concrete Construction Engineering Handbook. Edited by Edward G. Nawy, CRC Press 2008.

[37] P. Shafigh, U. J. Alengaram, H. B. Mahmud, M. Z. Jumaat. 2013. Engineering properties of oil palm shell lightweight concrete containing fly ash. *Materials and Design* **49**, pp. 617-621.

[38] M. M. U. Islam, K. H. Mo, U. J. Alengaram. 2016. Durability properties of sustainable concrete containing high volume palm oil waste materials. *Journal of Cleaner Production*. **137**. pp. 167-177.

[39] K. H. Mo, T. S Chin, U. J. Alengaram, M. Z Jumaat. 2016. Material and Structural properties of waste –oil palm shell concrete incorporating ground granulated blast-furnace slag reinforced with low-volume steel fibres. *Journal of Cleaner Production*. **133**. pp. 414-426.

Chapter 5

Engineering Properties of OPS Concrete

Keywords: OPS concrete, lightweight concrete, fresh concrete properties, hardened concrete properties, durability, cost efficiency

5.1. Introduction

Oil palm is one of the most important industries in Malaysia. This industry produces over 5 million tons of oil palm shell (OPS) solid waste annually. Converting the OPS into a useful new construction material, conserves natural resources and has great environmental benefits. The OPS aggregate being a renewable resource presents a potential alternative to the traditional concrete material, particularly an aggregate for lightweight concrete.

The numerous investigations conducted on OPS concrete confirm that the utilization of OPS as coarse aggregate in structural lightweight concrete production satisfies the technical requirements. OPS aggregate can effectively be used in making structural lightweight concrete. The utilization of OPS as coarse aggregate is a pioneer study. The OPS aggregate is beneficial to the construction industry particularly for rural low cost houses, plantation houses and precast products for non-structural applications, if OPS aggregate is available locally.

This chapter describes the wonderful investigation conducted on the basic engineering properties of OPS concrete over 34 years from 1982 to 2016. Based on the investigation, OPS concrete is called structural lightweight concrete having a dry density of less than 2,000kg/m^3 and 28-day compressive strength of 17-45N/mm^2. This chapter explains the fresh concrete properties, hardened concrete properties, and durability properties of OPS concrete. How these basic engineering properties are influenced with mix ingredients, mix proportions, compaction, curing, etc. are elaborated.

The fresh density, air content and slump of OPS concrete are explained in the fresh property section. The hardened concrete properties of OPS concrete are explained with hardened density, compressive strength, flexural strength, splitting tensile strength, modulus of elasticity, bond strength and shear strength. The durability properties of OPS concrete are elaborated with elevated temperature, drying shrinkage, water absorption, sorptivity, volume of permeable void, 90-day salt ponding, rapid chloride penetration test (RCPT) and sulphate resistance capacity. Finally, the cost efficiency of OPS concrete is discussed.

5.2. Fresh Properties of OPS Concrete

The fresh properties of concrete depend mainly on the ingredients, mix proportion and environmental conditions. Generally, aggregates occupy about 60-80% of the total volume of concrete. Total specific areas of aggregate are minimized through proper selection of size, shape and proportion of fine and coarse aggregates.

The shape and surface texture of aggregates control the void content and affect the water requirement of concrete mix significantly. The fineness modulus of aggregate is a numerical index of fineness giving some idea of the mean size of that aggregate. The fineness modulus of aggregate is a prime indication for most economic mix designs to obtain the required workability and strength.

The fresh concrete properties and bonds between aggregates and the mortar phase are influenced significantly by physical properties such as roughness, shape and texture of aggregates. The smoothness or roughness of aggregate reflects the surface texture. The glassy, smooth, granular, rough, crystalline, porous and honeycombed textures are the visual properties of the surface of aggregates [1].

5.2.1. Slump and Air Content

The slump as the workability measuring indicator is a crucial factor for fresh concrete. The slump is the most important factor for achieving the good workmanship of concrete for better strength and durability. Table 5.1 shows slump values of OPS lightweight concrete in a wide range from almost zero to a high of 160mm. The higher slump values were influenced by w/c ratio, superplasticiser, supplementary cementing material (SCM), pre-treatment of OPS aggregates etc. The w/c ratios were ranged from 0.30 to 0.80. The amount of binder used was in the range from 295 to 616 kg/m^3 concrete. Usually, lightweight concrete, particularly OPS lightweight concrete needs more binder as some binder may be left at the convex side of OPS aggregate without proper hydration. The SCM using slag used up to 70% binder replacement. The rate of slump loss is very much affected by the rate of the absorption of mix water by the lightweight aggregate and the ambient condition. The OPS aggregate with a saturated surface dry (ssd) condition has been used in mixing as OPS has a high water absorption capacity [2, 3, 4].

The OPS aggregates are lighter than other ingredients in the mix and they rise to the upper concrete layer if the compaction of fresh concrete is excessive. The author has noticed this situation during compaction of precast prototype floor panels of 3 to 8m long using OPS concrete as discussed in Chapter 7. Generally, slump should be limited to a maximum of 100 mm for OPS concrete.

Figure 5.1 shows the rough surface appearance of fresh OPS concrete while casting a prototype RC member of 1m wide and 3m long. The mix used was 1:1.4:0.76:0.70 (OPC:Sand:OPS:Granite) by the weight of cement including w/c ratio of 0.38. The cement content used was 450kg/m^3 and superplasticiser used was at 1.7litres/100kg cement. The slump, fresh density, 28-day air-dry density and 28-day compressive strength were 90mm, 2002kg/m^3, 1917kg/m^3 and 38.53N/mm^2 respectively. The duration of using poker hammer for the compaction was crucial to avoid accumulation of OPS aggregates on the surface. OPS aggregates are lightweight having a specific gravity of less than 1.5 compared with sand, granite stone and cement. While Figure 5.2 shows the flat surface appearance of fresh NWC using ready-mixed concrete for casting a prototype RC member of 1m wide and 3m long. The compaction was done without using a poker hammer for this NWC while maintaining both concretes, OPS concrete and NWC of almost the same grade.

Table 5.1. Mix proportions and slump values of OPS concrete (courtesy, author)

Reference and year	Mix design			Slump (mm)	Remarks
	w/c ratio	Cement (kg/m^3)	Mix proportion (C:sand:OPS)		
[6], 1988	0.48 0.54 0.65	335 314 295	1:1.70:2.08 1:1.88:2.18 1:2.10:1.12	8 28 50	<80mm slump, difficult in casting for prototype structural member
[7], 1990	0.5 0.6 0.7 0.8 0.5 0.6 0.7 0.8	-	1:1:2 1:2:4	30 63 collap. collap. 3 28 55 80	
[2], 2001	0.41	480	1:1.71:0.77	<10	
[3], 2007	0.38	510	1:1.66:0.60	50-70	
[8], 2006	0.41	-	1:1.71:0.77	90-95	Slump improved through PVA treated OPS
[9], 2009	0.35	596 616 585 540	1:1:0.80 1:0.91:0.73 1:1.09:0.72 1:1.45:0.73	160 103 65 60	Influence of SP
[4], 2010	0.38	450	1:1.40:1.45	90	30% OPS replacement with granite
[10], 2011	0.35	500	1:1.2:0.8	80-100	Influence of FA, SF and SP
[11], 2013	0.35	500	1:1.56:0.84	35-75	OPC replaced with slag at 0 to 70%
[12], 2015	0.34	-	1:1.68:1.24	30-90	Use of fibres, granite and SP
[13], 2016	0.30	565	1:1.7:0.65	~35	Cement replacement up to 25% with POFA
General notes	**0.30-0.80**	**295-616 kg/m^3, usually required more binder for LWC**	**-**	**80-160mm slump, Suitable for casting of prototype RC structure**	

Table 5.2. Air content in fresh OPS concrete [2]

Mix proportion (mix by weight of OPC, OPC=480kg/m^3, free w/c ratio=0.41)	Air content in fresh concrete (%)
1:1.71:0.77 with 0% OPC replacement with fly ash	4.9-5.1
1:1.71:0.77 with 10% OPC replacement with fly ash	4.9-5.0
1:1.71:0.77 with 15% OPC replacement with fly ash	4.8-4.9

Figure 5.1. Rough surface appearance of freshly cast prototype RC floor panel using OPS concrete with slump value of 90mm, compaction performed using poker hammer, [courtesy, author].

Figure 5.2. Flat surface appearance of freshly cast prototype RC floor panel using ready-mixed concrete (normal weight concrete) with slump values of 125mm, compaction performed without poker hammer [courtesy, author].

In fresh concrete, air content is relatively high in OPS concrete. Table 5.2 shows air content of OPS concrete in the range from 4.8 to 5.1% [2]. No superplasticiser was used in this mix. OPS aggregates come in many shapes, such as roughly parabolic or semi-circular, flaky and other irregular shapes and these are obstacles for achieving full compaction of OPS concrete. OPS aggregates are also porous materials as observed by water absorption, and these may cause high air entrapment in fresh concrete. In the United Kingdom, the total air content of 4.5±1.5% by volume of a concrete mix is recommended [5].

5.2.2. Fresh Concrete Density

Fresh concrete density is an early indication of the type of hardened concrete, whether it is ultra-lightweight or lightweight or structural lightweight or normal weight or heavy weight concrete. In this investigation, OPS concrete mixed without fly ash and with fly ash, has produced fresh density in the range of structural lightweight concrete (Table 5.3). Normally, the fresh concrete density for lightweight concrete is around 100 to 200 kg/m^3 higher than the 28-day air-dry density [14]. An air-dry density is relevant for the weight of concrete in a structure. The density of structural lightweight concrete is within the range of 1,200 to 2,000 kg/m^3 [14-17]. Fresh concrete densities of OPS concrete revealed by other researchers are in the range from 1,600 to 1,960 kg/m^3 [6-8, 18-21].

Table 5.3. Fresh concrete and 28-day air-dry densities of OPS concrete [2]

Mix proportion (mix by weight of OPC, OPC=480kg/m^3, free w/c ratio=0.41)	Fresh and 28-day air-dry densities (kg/m^3)	
	Fresh density	28-day air-dry density
1:1.71:0.77 with 0% OPC replacement with fly ash	1,958	1,856
1:1.71:0.77 with 10% OPC replacement with fly ash	1,923	1,817
1:1.71:0.77 with 15% OPC replacement with fly ash	1,910	1,801

5.3. Hardened Density of OPS Concrete

Concrete is a heterogeneous material whose properties depend on both the properties of the individual ingredients and their compatibility. The important properties of lightweight concrete [14, 17, 22, 23] are:

a) Density
b) Strength
c) Modulus of elasticity
d) Bond and anchorage
e) Impact resistance
f) Thermal conductivity
g) Performance at high temperature and
h) Acoustic behaviour

Generally, lightweight concrete is produced either for structural or thermal insulation in the building. The general properties for structural lightweight concrete are:

a) High strength with low density
b) High modulus of elasticity and
c) Low rate of shrinkage and creep

A low thermal conductivity is necessary for insulation purposes when lightweight concrete is used. The behaviour of lightweight concrete is closely related to its density. The principal factors influencing the density are [14]:

a) Cement content in mix
b) Relative density of aggregates
c) Entrained air
d) Moisture content of aggregates and
e) Environmental conditions

The high internal porosity of lightweight aggregate is a common feature, which is the main reason for their low bulk specific gravity. The dry density of compacted concrete made with different lightweight aggregates varies from about 300 to 2,100 kg/m^3 [14, 15, 17, 24]. The density of structural lightweight concrete is mentioned in many codes of practice. In Table 5.4, the densities of different kinds of lightweight and normal weight concretes, together with their application are presented [14, 15].

Lightweight concrete is identified as concrete having a density of 2,000 kg/m^3 or less [25, 26]. According to [27], the density classification of lightweight concrete is shown in Table 5.5. Lightweight concrete, whether made with natural or manufactured lightweight aggregate needs more binder than conventional normal weight concrete [28].

The densities of OPS concrete shown in Table 5.6 are in the range of 1,600 to 2,050 kg/m^3 depending upon the mix proportions, supplementary cementing materials (SCM), fibre, dosage of superplasticizer etc. [2, 6-7, 9, 11-12, 18-19, 29-33].

Table 5.4. Densities and applications of different types of concrete [14, 15]

Type	Concrete density (kg/m^3)	Application
Ultra lightweight	300-1,100	Insulation
Lightweight	1,100-1,600	Masonry
Structural lightweight	1,450-1,900	Structural
Normal weight	2,100-2,550	Structural
Heavyweight	2,900-6,100	Radiation

Table 5.5. Density classification of lightweight concrete according to [27]

Density Class	1.0	1.2	1.4	1.6	1.8	2.0
Oven-dry density (kg/m^3)	801 – 1,000	1,001 – 1,200	1,201 – 1,400	1,401 – 1,600	1,601 – 1,800	1,801 – 2,000
Density of plain concrete, (kg/m^3)	1,050	1,250	1,450	1,650	1,850	2,050
Density of reinforced concrete, (kg/m^3)	1,150	1,350	1,550	1,750	1,950	2,150

Table 5.6. Mix proportions and properties of OPS concrete [courtesy, author]

Reference and year	Mix design			Hardened density	28-day properties				Remarks
	w/c ratio	OPC (kg/m^3)	Mix proportion (OPC:sand:OPS)	Hardened density (kg/m^3)	Comp. St (N/mm^2)	Flexural St. (N/mm^2)	Splitting tensile St. (N/mm^2)	Modulus of elasticity, E (GPa)	
[29], 1982	-	-	-	1,750-1,915	21.1-30.0	4.5-4.7	-	5.5	
[30], 1984	0.6	-	1:1.5:0.5	2,050-1,725	2.5 - 10.0	-	-	-	Comp. St. less than 17MPa
[6], 1988	0.48 0.54 0.65	335 314 295	1:1.70:2.08 1:1.88:2.18 1:2.10:1.12	1,763-1,753	~23 ~22 ~16	6.20 5.50 4.30	2.40 2.35 2.00	-	
[7], 1990	0.5 0.6 0.7 0.8 0.5 0.6 0.7 0.8	-	1:1:2 1:2:4	1,780-1,630 1,700-1,600	22.20 19.80 16.50 14.90 18.90 16.50 13.00 11.50	2.81 2.53 2.30 2.13 -	-	-	
[2], 2001	0.41	480	1:1.71:0.77	1,905-1,890	19.50- 29.40	-	-	-	use of FA and $CaCl_2$
[34], 2002	0.41	-	1:1.71:0.77	-	20.12- 24.22	2.75-4.00	1.78-2.41	7-7.6	Different curings
[3], 2007	0.38	510	1:1.66:0.60	1,995-1,930	25-28	-	-	-	use of SP
[8], 2006	0.41	-	1:1.71:0.77	1,865-1,930	28.22- 32.84	-	-	-	Use of PVA solutions
[19], 2006	0.75 0.50	-	1:2:4 1:1:2	1,630 1,700	~15 15	-	-	-	OPS of 5-15mm size

Table 5.6. (Continued)

Reference and year	Mix design			Hardened density	28-day properties				Remarks
	w/c ratio	OPC (kg/m³)	Mix proportion (OPC:sand:OPS)	Hardened density (kg/m³)	Comp. St (N/mm²)	Flexural St. (N/mm²)	Splitting tensile St. (N/mm²)	Modulus of elasticity, E (GPa)	
[9], 2009	0.35	596 616 585 540	1:1:0.80 1:0.91:0.73 1:1.09:0.72 1:1.45:0.73	1,630 1,642 1,685 1,728	26.98 29.49 34.49 37.79	2.79 2.76 3.22 4.10	1.98 1.90 2.00 2.35	7.08 8.57 10.01 10.90	use of SP
[4], 2010	0.38	450	1:1.40:1.45	1,917	38.53	5.79	3.50	21.10	30% OPS replacement with granite, majority of OPS shell thickness of more than 3mm
[10], 2011	0.30	595 585 573 560 540 585 616 654 596	1:0.73:0.91 1:0.74:0.91 1:0.74:0.92 1:0.80:1.0 1:1.45:0.73 1:1.09:0.72 1:0.91:0.73 1:0.73:0.73 1:1:0.80	-	23.0-37.8	-	-	5.5-10.9	use of FA, SF and SP
[11], 2013	0.35	500	1:1.56:0.84	1,826-1,802	42.5-23.0	-	-	-	OPC replaced with slag at 0 to 70%, oven dry density, curing effect
[32], 2015	0.33	550	1:1.7:0.65	2,000-1,940	45.1-33.4	5.71-4.03	3.51-2.31	17.6-12.6	OPC replaced with slag at 0 to 70%
[33], 2015	*0.25	#b=463.41	1:2.5:0.60	1,930-1,600	36.0-25	4.30-2.40	2.41-1.40	14.65-5.0	Geopolymer concrete

Reference and year	Mix design			Hardened density	28-day properties				Remarks
	w/c ratio	OPC (kg/m^3)	Mix proportion (OPC:sand:OPS)	Hardened density (kg/m^3)	Comp. St (N/mm^2)	Flexural St. (N/mm^2)	Splitting tensile St. (N/mm^2)	Modulus of elasticity, E (GPa)	
[12], 2015	0.34	-	1:1.68:1.24	2,014-1,996	~45- ~40	~5.6-~4.8	~5-~2.75	-	Use of fibres, granite and SP
[13], 2016	0.30	565	1:1.7:0.65	-	42.4-32.0	6.5-3.0	4.0-2.6	15.4-11.4	Cement replacement up to 25% with POFA and curing effect
[35], 2016	0.30	560	1:1.8:0.63	-	~41-~32	-	~5.75-~3	~17-~10	Cement replacement up to 60% with Slag, and, use fibre and SP
General notes	**0.25-0.80**	**295-654, usually required more binder for LWC**	-	**1,600-2,050kg/m^3, indicating LWC**	**up to 45.1 N/mm^2, mostly above 17 N/mm^2**	**2.13-6.5**	**1.25-~5.75, mostly more than 2MPa**	**up to 21.10GPa, significant improvement**	**Recognized as new structural LWC**

* a/b ratio- alkaline activator to binder ratio=0.50, w/b =0.25.

#b-binder=Rice husk ash+slag+metakaolin (1:2:1).

5.4. Influence of Curing on Strength Development

In most laboratory conditions, concrete is cured in potable water for up to 28 days [36]. At construction sites, most curing techniques used currently have practical drawbacks. Plastic wrap around the concrete is seen to be a good curing method. Plastic film curing [23] is considered to minimize moisture loss from the surface of concrete where continuous spraying of water is not possible especially during the early stages of curing. Damp hessian is seldom kept damp and can act as a wick to withdraw water from concrete. A concrete surface ponded with water is the closest method to a laboratory curing method. The spraying of water on the concrete surface at site is a usual method adopted in the field.

It is recognized that ambient temperature, particularly the curing temperature, has the major effect on the engineering property of concrete, particularly the compressive strength development [37]. The higher the ambient temperature, the higher the early strength in the first stage of strength development of concrete because higher temperatures speed up the chemical reactions in hydration. The CSH gels under intensifying physical or chemical factors produce a higher specific surface at the early ages and by this, the finer texture and lower porosity at early stages produce higher strength [38]. Generally, the best curing temperature for most concretes is between 15 and 40^{0}C [39]. The ambient temperature noticed in university laboratories in Malaysia, is between 24-28^{0}C.

The strength of concrete develops with the progress of hydration. When the hydration process stops, the strength development also stops. The hydration stops when there is no more unhydrated cement available in concrete. Not enough free water means diffusion can no longer take place [37]. The effect of water has a considerable influence on the engineering properties of hardened concrete because it keeps the concrete wet. Curing at early stages obstructs cracking. The hydration continues with the presence of water, and free water can influence the strength of concrete [40].

5.5. Compressive Strength of OPS Concrete

The compressive strength is the most important property of concrete for designing concrete structures. It has become common practice to assess concrete quality with its compressive strength. The compressive strength test is easy and relatively inexpensive. It is assumed that most of the important properties of concrete such as the modulus of rupture, splitting tensile strength and modulus of elasticity are directly related (at least qualitatively) to the compressive strength [15, 24, 41]. Since concrete has very low tensile strength and is used primarily in a compressive mode for engineering practice. The structural design codes are based mainly on the compressive strength of concrete.

The strength development of lightweight concrete depends on the mortar strength, coarse aggregate strength and strength of adhesion between coarse aggregates and mortar. Chen et al. (1999) [42] have also observed in their study that both particle density and particle sizes of lightweight aggregate are controlling factors for the strength.

Porosity affects compressive strength. Generally, the higher the porosity, the lower the strength. Factors affecting compressive strength include:

a) Strength and stiffness of aggregate particles
b) Water/cement ratio (w/c)
c) Binder (cement) content in the matrix
d) Curing age and
e) Density of aggregates

As in the case of normal weight concrete, w/c ratio has the same effect on the compressive strength of lightweight concrete. Due to absorption of water by lightweight aggregate, it is difficult to measure the reduction of the effective w/c ratio [14]. The relation between compressive strength and w/c ratio is different for every type of lightweight aggregate and should be taken into account separately [17]. The strength increases with cement content for a given workability and it depends also on the aggregate type.

Lightweight concrete (LWC) is cost-saving for commonly required compressive strengths in the precast industry [43]. The common design compressive strengths are of 21 to 35 N/mm^2 and it can be above 35 N/mm^2 in precast plants [44]. Lightweight aggregates like other conventional normal weight aggregates have strength ceilings and this can be increased by reducing the size of coarse aggregate.

The 28-day compressive strengths of OPS concrete as shown in Table 5.6 are mostly in structural grade having more than 17N/mm^2 [45]. The strength development is influenced by several factors such as w/c ratio, mix proportion, binder content, curing, supplementary cementing materials (SCM), fibre, geopolymer, OPS treatment with PVA solution etc.

Two factors that control the development of strength in concrete are interparticle bond which offers the origin of strength and the porosity of paste which is a strength reduction factor [37]. The compressive strength is also controlled by the strength of the aggregate phase as well as the strength of the mortar phase, and the compressive strength of specimens depends on which of these two fails first. Lightweight aggregate, however, requires a higher total water content to compensate for the absorption of water by aggregates. The strength, stiffness and density of OPS aggregate are governing factors for the compressive strength in concrete. The surface of OPS aggregate is smooth. Hence, interparticle bond by OPS aggregate is less than with conventional crushed aggregate which plays a vital role in strength development.

At early stages, the compression failure pattern in OPS concrete as observed from the broken cubes has indicated that failure is due to the breakdown in the bond between the mortar phase and OPS aggregates. This has indicated that adhesion between OPS and the mortar phase is not strong because of the smooth surface of OPS aggregate. But in the later stages, the bond between the mortar phase and the OPS aggregate phase becomes stronger as the crack path travels through OPS aggregate [2, 31].

Generally, the lower the density of concrete, the lower the concrete strength [23]. OPS concrete is lighter than normal weight concrete. Porosity also reduces the compressive strength. OPS concrete has relatively high porosity as observed by the air content in fresh OPS concrete [2]. The strength, stiffness, thickness and density of OPS aggregate are also low and these are the governing factors for the compressive strength in concrete. OPS aggregates are of many shapes such as roughly parabolic, circular or semi-circular, flaky and other irregular shapes, and are also controlling factors for strength development [42]. Strength and density depend on the type of aggregate, the richness of the mix and compaction [23].

In general, lightweight concrete requires more binder (cement) content than normal weight concrete [14]. The researchers for OPS concrete have used binder content in the range of 295 to 654 kg/m^3 concrete (Table 5.6). Based on 28-day compressive strength as shown in Table 5.6, OPS concrete is recommended as structural lightweight concrete for having compressive strength in the range of 17.5 to 45.1 N/mm^2. The compressive strength of more than 35 MPa is suitable for precast floor which fulfills the strength requirement for service conditions [46, 47]. There are five different requirements for the capacity determination while using a precast floor system such as (i) bearing capacity, (ii) shear capacity, (iii) flexural capacity, (iv) deflection limit, and (v) handling restrictions [48]. Based on compressive strength, it can be concluded that OPS concrete is suitable for structural applications, particularly rural low cost houses, plantation houses, etc. As non-structural applications, this concrete can be used as a precast component for drain systems, retaining walls, slope surface protection, road curbs etc.

Table 5.7 shows the 28-day compressive strengths of different types of concrete including applications [14, 15]. Table 5.8 shows the air-dry density and compressive strength of different lightweight concrete other than OPS concrete. Figure 5.3 shows an insight of OPS concrete after splitting tensile tests using concrete cylinders. It shows good bonding between the mortar phase and OPS aggregates.

Table 5.7. 28-day compressive strength of different types of concrete [14, 15]

Type	28-day Comp. strength (N/mm^2)	Application
Ultra lightweight	>0.5	Insulation
Lightweight	7-14	Masonry
Structural lightweight	>15	Structural
Normal weight	20-40	Structural
Heavyweight	20-40	Radiation

Table 5.8. Comparison of air-dry density and 28-day compressive strength of different lightweight concretes other than OPS concrete [courtesy, author]

Type of lightweight concrete	Concrete density (kg/m^3)	28-day Comp. st. (N/mm^2)	Reference
Clay-Sewage sludge concrete	1,410-1,500	15.9-17.0	[49]
Cork granule concrete	890	12.0	[50]
Diatomite concrete	992	11.2	[17]
Expanded Clay concrete	1,760	62.0	[17]
Expanded Perlite concrete	640	1.8	[51]
Expanded Slate concrete	1,360	27.6	[17]
Foamed slag concrete	1,760	22.7	[17]
OPBC concrete	1,845-1,980	27.0-35.0	[52]
Pumice concrete	1,450	29.0	[51]
Sawdust concrete	1,600	34.5	[17]
Sintered PFA concrete	1,745	45.0	[17]
Vermiculite concrete	560	1.2	[51]
Water treatment sludge concrete	1,694-1,789	24.8-43.2	[53]

Figure 5.3. OPS concrete- an insight after splitting tensile test at later age [courtesy, author].

5.6. Flexural Strength of OPS Concrete

The flexural strength or modulus of the rupture test is very important for the design of concrete road slabs and airfield runways because the flexural tension is a critical factor [23, 24,54]. The correct choice of the aggregate/cement ratio is, in most cases, of greater importance than the choice of the water/cement ratio. The modulus of rupture for the design of concrete slabs is normally in the range of 4.15 to 4.85 N/mm^2 [55]. The modulus of rupture is usually about 15% of the compressive strength for normal weight concrete (NWC). It is an important value to estimate the load magnitude at which cracking will develop. The flexural strength of LWC is in the same order with NWC at early stages [1].

The flexural strength increases with the density of concrete but it is relatively less in comparison with the compressive strength increase. Flexural strength, in practice, is more commonly used in beams and slabs, and is very important to avoid cracking, though its magnitude is low compared to compressive strength. Flexural strength, as with compressive strength is also dependent on the strength of coarse aggregate. The flexural strength is largely influenced by the diffused moisture distribution in the test samples [17].

Flexural strength is very sensitive to the moisture content and distribution in the sample. Drying for only 30 minutes can cause a reduction of 8% in the measured flexural strength [37]. The approximate relationship between flexural strength and compressive strength for structural lightweight concrete and normal weight concrete is 0.09 to 0.12 and 0.1 to 0.30, respectively [37]. Based on results shown in Table 5.6, generally, OPS concrete has shown an approximate relation between flexural and compressive strength around 0.10 to 0.15.

Mahmud et al. (2009) [9], based on their investigation have developed a relation between flexural and compressive strengths of OPS concrete as shown below with regression, R=0.92.

$$f_{flex}= 0.3(f_c)^{2/3} \quad \text{(Eq. 5.1)}$$

where,

f_{flex}= flexural strength (N/mm^2)

f_c = compressive strength (N/mm^2)

5.7. Splitting Tensile Strength of OPS Concrete

The tensile strength of concrete is an important property in structural design since it affects bond strength and flexural cracking. The split tensile test gives an indication of the bond between aggregate and concrete [28]. No direct tensile test for concrete is mentioned in the standards and it explains the procedure to carry out an indirect tensile test by compressing a concrete cylinder. This test can provide an estimation of the tensile strength of concrete.

The porosity and strength of aggregate particles affects the strength of concrete. The tensile strength of an aggregate fraction is the average value of the tensile strength of the particles. Aggregates with densities of 1,400 to 1,500 kg/m^3, can develop maximum particle tensile strengths of 2.3 to 4.6 N/mm^2 [22]. As with flexural strength, the splitting tensile strength is dependent on the curing condition.

In a study on OPS concrete, the ratio of splitting tensile strength to compressive strength is found to be almost in the same order as that of normal weight concrete (NWC as control concrete where coarse aggregate replacement is identical to OPS aggregate) as shown in Table 5.9 [2]. The relationship between splitting tensile strength and compressive strength of lightweight concretes [17] such as Aglite, foamed slag, Leca and Lytag lightweight concrete is found in the following way:

$$f_{ct} = K\sqrt{f_c} \quad \text{(Eq. 5.2)}$$

where,

f_{ct} = cylinder splitting tensile strength (N/mm^2)

f_c = cube compressive strength (N/mm^2)

K = coefficient

For the unit of N/mm^2, the K values of the above conventional lightweight concretes are found to vary from about 0.37-0.5 [17]. In the observation, the K values of OPS concrete and NWC as control concretes are shown in Table 5.10 [2]. The K values of OPS concrete for 28-day onwards are between 0.38 and 0.62, depending upon mix orders and curing conditions. For 28 days onwards, the K values of NWC are between 0.53 and 0.74 depending also on mix orders and curing conditions. The K values of OPS concrete are relatively lower than NWC.

Concrete structures are subjected to tensile stresses due to drying or temperature changes. Factors influencing compressive strength also influence tensile strength [14] and usually, the tensile strength of concrete is considered to be about one-tenth of its compressive strength. As compared to compressive strength, the tensile strength is more affected by improper curing. Improper curing is responsible for the formation of inferior quality gel and shrinkage cracks [51]. For most lightweight concretes, the splitting tensile strength is less than that of normal weight concrete of the same compressive strength [17].

Generally, splitting tensile strength is about 60% of the modulus of rupture [17]. The ratios between splitting tensile and flexural strength of both OPS concrete and NWC as control concrete are shown in Table 5.11 [2]. The ranges of ratios for OPS concrete and NWC of different mix orders under two curing conditions are 0.48-0.92 and 0.71-0.83, respectively. A wide range in the variation of the ratios of splitting tensile- flexural strength, (of 0.39-0.91) is reported elsewhere [37]. The tensile strength is also largely affected by moisture content. The shell thickness of OPS aggregate plays a significant role in splitting tensile strength development [4]. OPS concrete under full water curing shows higher splitting tensile strength than that of partial water curing conditions. OPS concrete has shown the relation between splitting tensile and compressive strength of around 0.06 to 0.11 (Table 5.6).

Table 5.9. Ratio of splitting tensile to compressive strength of OPS concrete [2]

Type of concrete	*Curing	Ratio of splitting tensile to compressive st.				
		3-day	7-day	28-day	56-day	90-day
OPS concrete with 0% fly ash	Partial water	0.08	0.07	0.08	0.09	0.09
	Full water	0.08	0.07	0.09	0.12	0.11
OPS concrete with 10% fly ash	Partial water	0.08	0.07	0.09	0.09	0.09
	Full water	0.08	0.07	0.09	0.12	0.11
NWC with 0% fly ash	Partial water	0.04	0.04	0.04	0.04	0.04
	Full water	0.04	0.04	0.06	0.06	0.05
NWC with 10% fly ash	Partial water	0.11	0.09	0.08	0.09	0.08
	Full water	0.11	0.09	0.10	0.09	0.09

*Note:
Partial water curing= 1-day in mould followed by 6 days in water and rest days in room until test
Full water curing= 1-day in mould and rest days in water until test.

Table 5.10 K values of relation between splitting tensile strength and compressive strength of OPS concrete and NWC as control concrete according to Equation 5.1 [17]

Type of concrete	*Curing	K-values				
		3-day	7-day	28-day	56-day	90-day
OPS concrete with 0% fly ash	Partial water	0.30	0.32	0.39	0.44	0.45
	Full water	0.30	0.32	0.48	0.62	0.62
OPS concrete with 10% fly ash	Partial water	0.28	0.29	0.40	0.44	0.44
	Full water	0.28	0.29	0.47	0.57	0.58
NWC with 0% fly ash	Partial water	0.57	0.66	0.63	0.62	0.57
	Full water	0.57	0.66	0.73	0.74	0.67
NWC with 10% fly ash	Partial water	0.57	0.59	0.59	0.62	0.60
	Full water	0.57	0.59	0.70	0.69	0.69

*Note:
Partial water curing= 1-day in mould followed by 6 days in water and rest days in room until test
Full water curing= 1-day in mould and rest days in water until test

NWC- normal weight concrete.

Although OPS aggregate has a fairly smooth surface texture, it is seen that the ratio of splitting tensile strength to compressive strength of OPS concrete is comparable to NWC [2]. It is also found that the splitting tensile strength of OPS concrete is comparable to LWC made with Leca, Liapor, Perlite and Lytag by partial replacement to normal weight aggregate [4, 56].

Table 5.11. Ratio of splitting tensile strength to flexural strength [2]

Type of concrete	*Curing	Ratio of splitting tensile to flexural strength				
		3-day	7-day	28-day	56-day	90-day
OPS concrete with 0% fly ash	Partial water	0.86	0.52	0.64	0.74	0.74
	Full water	0.86	0.52	0.60	0.76	0.76
OPS concrete with 10% fly ash	Partial water	0.92	0.68	0.80	0.65	0.84
	Full water	0.92	0.68	0.76	0.90	0.91
NWC with 0% fly ash	Partial water	0.74	0.78	0.79	0.79	0.80
	Full water	0.74	0.78	0.75	0.80	0.79
NWC with 10% fly ash	Partial water	0.73	0.77	0.79	0.83	0.82
	Full water	0.73	0.77	0.75	0.79	0.80

*Note:
Partial water curing= 1-day in mould followed by 6 days in water and rest days in room until test
Full water curing= 1-day in mould and rest days in water until test.

5.8. Modulus of Elasticity of OPS Concrete

The density, pore structure and irregular texture of aggregates define the modulus of elasticity of aggregates. There is no standard method of test to obtain the modulus of elasticity for aggregate particles. The modulus of elasticity can be determined from the results of an ultrasonic test on single particles.

The modulus of elasticity of the aggregate affects the strength of the lightweight concrete. Generally, the higher the modulus of elasticity of aggregate, the higher the strength and stiffness of concrete. The moduli of elasticity of expanded clay and shale are in the range of 0.5 to 1.5 GPa [22]. There is no report available on the modulus of elasticity for OPS aggregate.

The lower modulus of elasticity, E, for lightweight concrete is liable for the deformation of structural members under specific loads. The modulus of elasticity of lightweight concrete is approximately 0.5 to 0.75 times that of normal weight concrete with the same compressive strength [16, 17]. The E-value of concrete depends on the stiffness of the constituent materials and their proportioning in the mix. The E-value decreases with (i) decrease in the stiffness of mortar, which depends on the proportion of binder, water and fine aggregates and the type of fine aggregate and (ii) decrease in the stiffness of coarse aggregate [14]. Therefore, under dynamic conditions, such as impact or load fluctuation, this lower stiffness can be beneficial for the structure.

Concrete's elastic property is influenced by the elastic properties of the ingredients and by the nature of the interfacial zone between aggregates and paste [57-59]. Due to the large volume of aggregates and their inherent stiffness, aggregate displays the major influence on the elastic modulus of concrete. The mineral admixture or cement type also influences the interfacial zone and bond in concrete.

The elastic behaviour of concrete depends on the bond strength and density of the interfacial zone, and on densities and void contents of concrete. The hydrophobic or hydrophilic nature of aggregate may contribute to the size of the interfacial region and the bonding of aggregate to the matrix. The interfacial regions become filled with hydration products.

In Salam's (1982) [29] investigation, the values of modulus of elasticity, E for OPS concrete is 5.5GPa. The development of the modulus of elasticity is influenced by the type of coarse aggregate, type of cement, w/c ratio of the mix and curing age. Alexander et al. (1995) [58], found that the E values of normal weight concrete using granite aggregate varied from 2.90 - 3.80 GPa for over 28 - 180 days, respectively. For a given w/c ratio, the rate of increment of elastic modulus depends on the type of aggregate and cement [58]. In a particular type of concrete, there is an optimum w/c ratio for the elastic modulus beyond which the stiffness shows little or no gain with the further reducing of w/c.

Table 5.6 shows significant improvement on the modulus of elasticity, E from 5.5 to 21.1 GPa of OPS concrete which is considered the breakthrough investigation of the 34 years from 1982 until 2016. The improvement on E is contributed by mix ingredients, binder, fiber, geopolymer, curing, etc. The E value is very essential input for structural design. Sharmin et al. (2015) [33], have developed a relation between the static modulus of elasticity (E_s) and compressive strength (f_c) on their geoplymer lightweight concrete using OPS and OPBC aggregate as shown in Equation 5.3 with R^2 of 0.94. Whereas, Alengaran et al. (2011) [10], have also developed the relation between static (E_s) and the dynamic modulus of elasticity (E_d) of OPS concrete as shown in Equation 5.4 with R^2 of 0.88.

$$E_s = 0.0036\sqrt{f_c}^{4.5793} \qquad \text{(Eq. 5.3)}$$

where,

E_s- modulus of elasticity (GPa)

f_c-compressive strength (N/mm^2)

$$E_s = 0.30(E_d)^{1.25} \qquad \text{(Eq. 5.4)}$$

where,

E_s- modulus of elasticity (kN/mm^2)

E_d-dynamic modulus of elasticity (kN/mm^2)

5.9. Structural Bond Properties of OPS Concrete

The structural bond strength is an important structural property of concrete. OPS concrete has shown 28-day compressive strength of more than 17MPa. The bond strength of OPS concrete through pull-out tests have been investigated [3, 21, 60, 61]. Universal Testing Machine (Shimadzu brand of 300 kN capacity) was used to carry out the pull-out test as the modified loading frame shows in Figure 5.4 [3]. According to ASTM C234 [62], the load was applied on the top of the concrete surface at a uniform rate without shock. The dial gauge was positioned at the unloaded end, as the free end of the concrete sample and the dial gauge readings were noted during subsequent loading at regular load intervals.

Table 5.12 shows the 28-day bond strength of OPS concrete cured in water using both round and deformed bars of 10, 12 and 16 mm diameter. The developed bond strengths were in the range of 3.0 to 9.36N/mm^2 with higher safety factors of 2.03 to 3.53 compared with BS8110, Part 1 [63]. Based on investigations, round bars (plain bars) failed by pulling out of the concrete without any visible cracking of the concrete cover. As for deformed bars, all samples failed by cracking the concrete cover. The bond strength increases with the decrease in the bar diameter. The bond is the development of anchorage and, it depends on the roughness and porosity of the surface of aggregate. The surface pores help in the development of good bonds by the suction of the paste into the pores. OPS aggregate in convex surfaces has pores of 16-24 micrometers [10] which is also a contributing factor in enhancing the bond strength. Hence, the bond strength behaviour of OPS concrete was very adequate for structural design and was similar to other lightweight concretes [3, 21, 60, 61] and was similar to sawdust lightweight concrete [64] and Lytag lightweight concrete [65].

Mo et al. (2016) [35], based on investigations have developed a relation between bond strength and compressive strength of OPS concrete as shown in the equations below. In their investigation for 1m^3 concrete, 520kg binder, 940kg sand, 400kg OPS and 170 kg water have been used including partial replacement of binder with 20% and 60% slag. They have used 12mm diameter steel bars for bond test.

$$\tau_{max}=3.5\sqrt{f_c} \quad \text{(Eq. 5.5)}$$

$$\tau= \tau_{max}\,(\delta/\delta_1)^{0.35} \quad \text{(Eq. 5.6)}$$

$$\tau_{fric} =0.25\tau_{max} \quad \text{(Eq. 5.7)}$$

where,

τ_{max}= maximum bond strength (MPa)
f_c = cylinder compressive strength (MPa)
δ_1 = slip corresponding to maximum bond strength (considered as 1.0mm)
τ_{fric} = residual bond strength (MPa)

Table 5.12. 28-day experimental bond strength of OPS concrete and safety factor over theoretical bond strength [66]

28-day comp. st. in water curing =28.1 (N/mm^2)	Theoretical bond st. for round bar (R type bar) using BS8110, Part 1 [63] = 1.48 N/mm^2			Theoretical bond st. for deformed bar (T type bar) using BS8110, Part 1 [63] = 2.65 N/mm^2		
Bar dia. (mm)	10	12	16	10	12	16
28-day experimental bond st. in water curing, (N/mm^2)	4.63	4.38	3.00	9.36	8.73	7.18
Safety factor = Exp. bond ÷ Theoretical bond st.	3.13	2.96	2.03	3.53	3.29	2.71

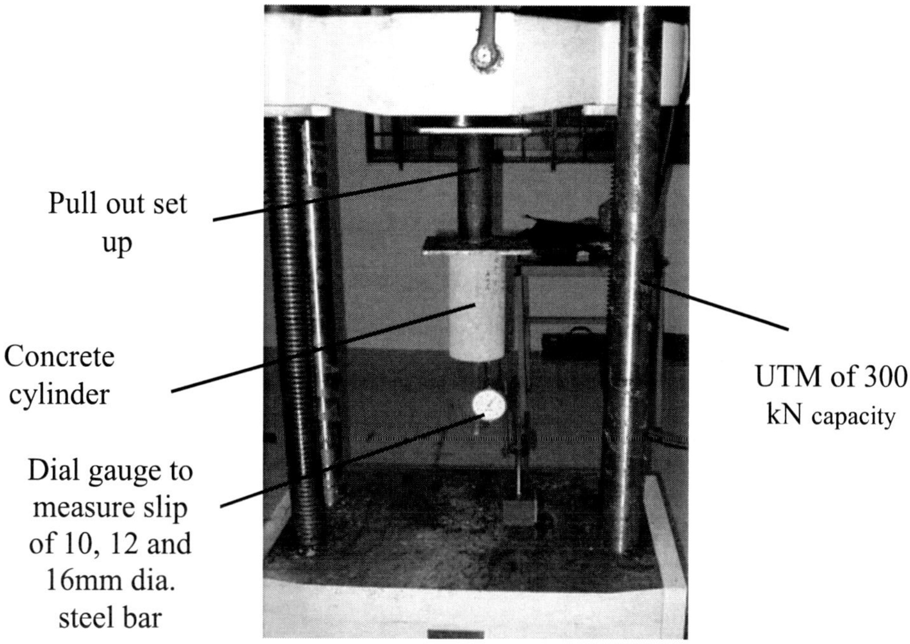

Figure 5.4. Bond test set up using 100mm dia. × 200mm long OPS concrete cylinder [66].

5.10. Shear Strength of OPS Concrete

The shear strength of the structural RC member is a very important property. Shear cracks occur in the region of large shear forces. Alengaram et al. (2011) [10], have investigated the shear strength using RC beams made with OPS concrete of grade 30. They have used beams of 150mm wide, 250mm deep and 2,600mm long with and without stirrups. The average shear strength of OPS concrete having concrete density of 1,850kg/m^3 is found 24% larger than conventional normal weight concrete. The different shapes, sizes and thicknesses of OPS aggregates have contributed to good interlocking of OPS concrete for improving the shear strength.

5.11. Durability Properties of OPS Concrete

The durability for building materials may be defined as the ability to resist adverse external influences such as climatic conditions, fire, chemical attack and mechanical damage. The principal conditions for the breakdown mechanisms in concrete are access to water and the permeability of the structure, which defines how fast aggressive liquids or ions are transported into or through the material.

The durability of concrete is related to its density [67]. Cement paste with higher density has greater resistance to the penetration of harmful agents into concrete. The density of cement paste depends on the pore structure. The pore size can be decreased using the ultrafine size of supplementary cementing materials (SCM) like fly ash (FA), palm oil fuel ash (POFA), silika fume (SF), slag etc.

As a porous media, cement and concrete materials adopt the pore structure with broad pore size distribution from nanometer to micrometer scales. Generally, the pores of concrete are classified into four major types:

a) Gel pores (< 10 nm diameter),
b) Transitional pores (10–100 nm diameter),
c) Capillary pores (100–1,000 nm diameter) and
d) Macropores (> 1,000 nm diameter)

Pores with different diameters have different influences on concrete. Pores whose diameter of 100 nm is defined as threshold. Pores of larger threshold sizes are assumed to be detrimental to concrete durability.

Generally, the concrete durability is related to abrasion resistance, carbonation, chemical resistance, fire resistance, frost resistance, permeability and water absorption [14, 68]. The durability of concrete can be divided into physical and chemical causes. The physical durability is a measure of volume change due to wetting and drying cycles, and the freezing and thawing cycles. The chemical durability results from the reaction between reactive silica in aggregates and alkalis contained in cement.

The rate of concrete deterioration is affected by the physical properties more than chemical composition, though the latter is known to be very important [69]. It depends largely on the permeability of concrete which is affected by the water/cement ratio, the quantity and particle size distribution of the cement and the incorporation of mineral admixture. Generally, lightweight concrete has more pores and has no special resistance. Due to this, lightweight concrete cannot be considered for use below the damp proof course.

5.11.1. Effect at Elevated Temperatures

In common practice, the design of reinforced concrete is based on normal operating conditions. Design consideration for the RC structure is also given to possible exposure to elevated temperatures during a fire accident. Fire is a natural hazard that destroys building properties. During fires, the temperature in concrete can be high.

The integrity of concrete during fires depends on four stages of heating cycles [70]:

a. First heating (rising temperature)
b. Constant temperature (peak temperature)
c. Cooling (dropping temperature) and
d. Post cooling (return to room temperature)

In non-load bearing masonry walls, the fire resistance rating varies from about 3 to 6 hours [17] depending on the thickness and constituent materials. As a common rule, a 100mm thick concrete block can resist fire for 2 hours if load-bearing, or up to 4 hours if non-load-bearing [71].

The oven temperatures at 100^{0}C, 200^{0}C and 300^{0}C for 2 hours (constant temperature) were selected to observe the performance of 100mm cubes of both OPS concrete and normal weight concrete (as control concrete) [2]. The time interval between the end of heating and the time of test was 36±4 hours, followed by returning the samples to room temperature. The purpose of this investigation was to study the concrete quality through density and compressive strength after heating with different specified temperatures as shown above. The results were compared with concrete cubes kept at room temperature (26^{0}C) as non-heated samples.

The significant reduction in the density of OPS concrete was observed when heated to high temperatures, up to 300^{0}C. The densities of OPS concrete subjected to this temperature varied from 94% for 100^{0}C to 85% for 300^{0}C compared to the density at non-heated condition (room temp. at 26^{0}C). The density of OPS concrete and control concrete at room temp. (26^{0}C) was 1,856kg/m^{3} and 2,300kg/m^{3} at age of 180-days respectively. Similarly, the densities of normal weight concrete (as control concrete) varied from 99% for 100^{0}C to 94% for 300^{0}C respectively compared with concrete cubes kept at room condition (26^{0}C).

The OPS aggregate is of organic origin and is a combustible material. Usually for wood material, burning occurs when it is heated up to about 250^{0}C, and escaping gases burst into flames [71]. At higher temperatures, particularly at 300^{0}C, the OPS aggregate in the upper layer of concrete turned black.

The compressive strengths of OPS concrete and control concrete at room temp. were 28N/mm^{2} and 56N/mm^{2} at 180-days respectively. Compared to compressive strength of non heated cubes at room condition (26^{0}C), the compressive strength of OPS concrete was about 63% at 100^{0}C, 45% at 200^{0}C and 33% at 300^{0}C. Similarly, in the control concrete, the compressive strength was 91% at 100^{0}C, 90% at 200^{0}C and, 94% at 300^{0}C. The compressive strength in control concrete at 300^{0}C was observed as higher than that at 100^{0}C and 200^{0}C treated samples. This was probably due to the hydration of unhydrated cement at a temperature of 300^{0}C.

Due to the high temperature and flow resistance, steam develops high pressure in the cement paste. Internal autoclaving is formed within the temperature range of 100 - 300^{0}C [72] because, within this range of temperature, steam is discharged most vigorously. In the internal autoclaving, the additional hydration of unhydrated cement is a result of the steam effect. During exposure to fire, the rise in temperature causes differential expansion between the cement matrix and the aggregates. This causes internal cracking in concrete and a loss of strength. Factors causing the degradation of concrete include the characteristics of concrete, temperature and loading conditions during fire exposure [71].

Aggregate plays an important role in concrete for the strength. The strength, stiffness and density of aggregate are important factors for the contribution to the compressive strength of concrete and these properties of OPS aggregates were affected by the elevated temperature. After a compressive strength test of the concrete sample, a closer examination was done on both OPS concrete and control concrete. It was seen that the reduction in the density, stiffness and strength of OPS aggregate was higher than the crushed stone aggregate at the elevated temperature.

In a report by [70], the compressive strength of concrete at 150^{0}C yielded results from as low as 30% to as high as 120% of the initial strength. Khoury (1992) [70], also mentioned that a drop in the compressive strength of concrete was seen at a temperature of 120^{0}C and a rise in the strength was visible at temperature 200^{0}C.

Mo et al. (2016) [35], investigated the 90-day old OPS concrete (mix proportion of binder: fine aggregate: OPS at 1:1.7: 0.65 with w/b ratio of 0.33 and binder replacement with slag at 0-60%, binder=550kg/m^3) at an elevated temperature, 200±5^{0}C for 4 hours. They observed the residual compressive strength of this concrete was between 17.7 and 15.7 N/mm^2 with a reduction of strength about 55-65% compared with non-heated concrete cubes.

Based on the above findings in elevated temperatures, OPS concrete as structural concrete is not advisable in fire exposure because of a noticeable reduction in the compressive strength unless precaution measures are taken. Hence, it needs further investigation on how OPS concrete can fulfill the structural requirement at elevated temperatures.

5.11.2. Drying Shrinkage

Shrinkage is the very important property of cement matrix and concrete. If water saturated concrete is exposed to a dry atmosphere, the surface of the concrete starts to lose water. Removal of water from concrete results in drying shrinkage. Aggregates have the preventive effect on the volume change happening in cement paste.

Shrinkage for concrete occurs due to the loss of free water in plastic concrete, by the settlement of solids, chemical reaction of cement paste and the drying of concrete. Similar to hydration of cement, drying shrinkage is a long lasting process, when concrete is exposed to drying conditions. For hardened concrete, drying shrinkage is a more important factor than other types of shrinkage such as plastic shrinkage, autogeneous shrinkage and carbonation shrinkage.

Several factors govern the drying shrinkage of lightweight concrete such as w/c ratio, degree of hydration, curing condition, cement composition, moisture content, admixture, aggregate (stiffness, content and volume/surface ratio), relative humidity, rate and duration of drying [15, 73].

The seasonal influencing factors that are responsible for shrinkage of concrete are rain, wind and solar radiation. The fineness of gel influences the drying shrinkage more. Cement paste shrinks more than mortar and mortar shrinks more than concrete. Concrete with smaller sized aggregates shrinks more than concrete with bigger sized aggregates [51].

Samples (100mm×100mm×500mm) of OPS concrete and normal weight concrete (as control concrete) were cured 1-day in the mould, 6 days in water and the rest of the days in the room during the drying shrinkage test [2]. At the stage of 7 days, samples were taken from the water tank for the setting of a gauge on the samples. The method mentioned by [74, 75] was followed with studs to the surface of concrete with epoxy adhesive. According to the configuration of the gauge (200 mm gauge length), two points were marked by a punch bar. After finishing the gauge settings, all samples were left inside the laboratory prior to testing at ambient condition. Drying shrinkage was measured using Demountable Mechanical (DEMEC) strain gauge with a digital read out. The accuracy of this gauge was about $\pm 5\times10^{-6}$ under the controlled laboratory condition. Results were reported at the stages of 7, 28, 56 and 90 days. Each result was the average of three samples.

Figure 5.5 presents the drying shrinkage results of OPS concrete, and control concrete using conventional crushed stone (OPS volume identically replaced with crushed stone) with different mixes using 0% and 10% fly ash (FA) as cement replacement, and 1.0% $CaCl_2$ (as additive). At 90-days, OPS concrete with 0% FA showed 14% higher shrinkage than control concrete mixed with 0%FA.

In this study [2], the drying shrinkage of OPS concrete at 28-days was 64 microstrain; 6% higher than control concrete. At a temperature of 20^0C and relative humidity of 65%, the final shrinkage of about 400 to 600 microstrain may be considered for lightweight concrete made with high quality expanded clay of 30-50 concrete grades [17, 22].

The 90 days drying shrinkage of OPS concrete was 185 microstrain and it was 14% higher than control concrete. According to [22], shrinkage of lightweight concrete is about 50% higher than normal weight concrete. The maximum drying shrinkage was found at the 90 days old OPS concrete containing 0% and 10% fly ash to be 185 microstrain and 213 microstrain respectively. Similarly, control concrete exhibited 160 microstrain and 149 microstrain. As compared with both concretes without calcium chloride, the 90 days old OPS and control concretes containing 1.0% calcium chloride produced the higher drying shrinkage of 243 and 215 microstrain respectively. The drying shrinkage was increased with drying age as observed here.

Mo et al. (2016) [35], investigated the 365-day old OPS concrete (mix proportion of binder: fine aggregate: OPS at 1:1.7: 0.65 with w/b ratio of 0.33 and binder replacement with slag at 0-60%, binder=550kg/m^3) for drying shrinkage. They maintained room ambient conditions (Temp. 20 ±2^0C, RH of 60±5%) during the study. The shrinkage values at 365-days were between 715 and 925 microstrain. The low stiffness of OPS shows less restraint against drying shrinkage as compared with conventional normal stone. Usually, LWC shows about 50% higher drying shrinkage than NWC [22]. The open texture and irregular surface of lightweight aggregate used in concrete making can show concrete final shrinkage of 1,000 microstrain [17].

Islam et al. (2016) [13], also studied drying shrinkage on OPS concrete up to the age of 64 days of drying. Their mix proportion was 1:1.7:0.65 by weight of cement (OPC:sand:OPS) with w/c ratio of 0.3 including 3.4 kg/m^3 superplasticiser. The binder was 565kg/m^3. They found the drying shrinkage of 350 microstrain at the age of 64 days. The drying shrinkage of concrete is influenced by the magnitude of porosity, size and continuous capillary system of hydrated cement paste in concrete [76].

5.11.3. Water Absorption

Yahaghi et al. (2015) [12], investigated water absorption of OPS concrete using 28-day old samples. They used a mix proportion of 1:1.68:1.24 by weight of cement (OPC: sand:OPS) with w/c ratio of 0.34 including fibres, superplasticiser and 60% granite as OPS replacement. The 28-day stage concrete showed the initial 30 minutes water absorption of about 2% and the final 3 days water absorption about 4.5% which were less than 3% and 10% by mass respectively [77].

Islam et al. (2016) [13], also studied water absorption on OPS concrete using 28-day and 90-day aged concrete. Their mix proportion was 1:1.7:0.65 by weight of cement (OPC:sand:OPS) with w/c ratios of 0.3 and 0.4 including superplasticiser and up to 50% OPC replacement with POFA. They found the initial 30 minutes water absorption of 2.2% and 6.2% for both ages and the final 3 days water absorption of 4.8% and 10.2% for both ages respectively. The influencing factors in reducing water absorption among others are curing age and POFA replacement. The pozzolanic POFA influenced to reduce pore size in concrete.

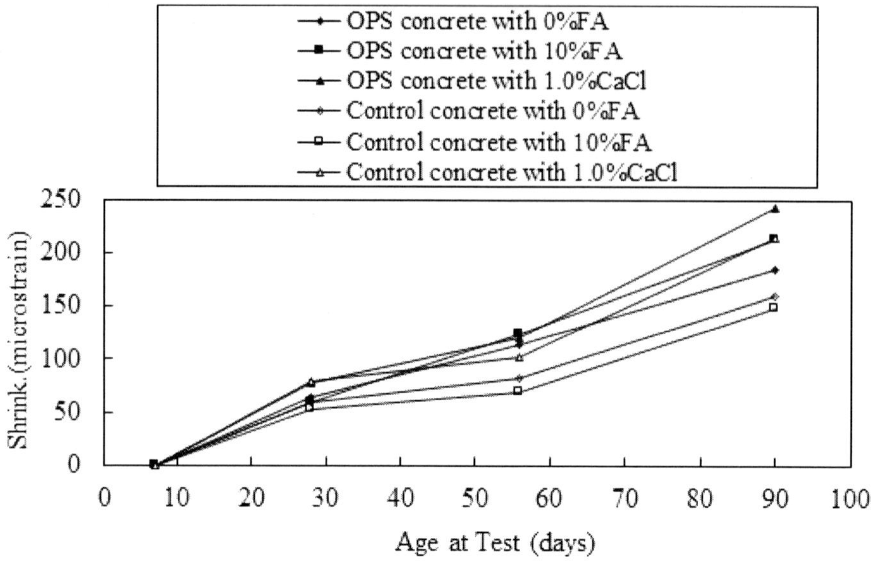

Figure 5.5. Drying shrinkage of OPS concrete and control concrete under full water curing condition (1-day in mould, 6 days in water and rest days in room ambient condition) [2].

5.11.4. Sorptivity

The concrete surface that is able to absorb and transmit water through capillary suction is known as sorptivity. This property is very crucial for the super-structure of RC building where rain water and salt ingress are controlled by absorption. The deleterious agent can enter for the deterioration of RC structure if the concrete surface is defective. This test shows the progressive transport of external agents from the concrete surface to interior concrete.

The sorptivity test shows a very useful indication of concrete pore structure [78]. Higher resistance of concrete towards water absorption can be seen with lower sorptivity values. Generally, concrete of high quality produces the sorptivity values of less than 0.1 mm/min$^{0.5}$ [79].

Teo et al. (2010) [80], investigated the sorptivity value of 28-day old OPS concrete using 100 mm cube specimens. They found the sorptivity values about 0.06 - 0.14 mm/min$^{0.5}$. Islam et al. (2016) [13], found a sorptivity value of 0.06 mm/min$^{0.5}$ using 28-day old OPS concrete. Their mix proportion was 1:1.7:0.65 by weight of cement (OPC:sand:OPS) with w/c ratio of 0.3 including superplasticiser. Mo et al. (2016) [35], also investigated this property for 28-day OPS concrete and the sorptivity value was 0.099 mm/min$^{0.5}$. Lightweight concretes made with Liapor (expanded shale) and Lytag (sintered pulverized fuel ash) lightweight aggregates have shown sorptivity values of about 0.03 and 0.06 mm/min$^{0.5}$ respectively [79]. The richness in mix, w/c ratio, compaction, curing, ssd condition of OPS etc. contributed to low sorptivity values of OPS concrete.

5.11.5. Volume of Permeable Voids

According to Shayan et al. (2006) [81], the volume of permeable void (VPV) is an important property of concrete as it affects the transport mechanisms through the concrete such as ingress of aggressive liquids and gases. Teo et al. (2010) [80], investigated VPV of OPS concrete according to [82]. The 28-day old OPS concrete showed VPV values about 20.1- 21.2%. The porous nature of OPS contributed to an increase in VPV.

The normal weight concrete has VPV values of about 12–16% [83, 84]. However, the results obtained from the study [81] were comparable to that of other lightweight aggregate concretes in which the VPV ranged from approximately 8.6% to 22.5% and [84].

5.11.6. 90-Day Salt Ponding

Teo et al. (2010) [80], investigated the 90-day salt ponding test on OPS concrete according to [85]. The coefficient of chloride diffusion was determined from the chloride profile by solving Fick's second law of diffusion. From the chloride profile, it was observed that the surface chloride concentration was high. This was due to the porous nature of OPS which allowed the deposition of chloride ions in its pores, as OPS has surface pores of 16-24 micrometers on a convex surface [10]. The combined accumulation of chlorides in OPS and cement paste pores therefore lead to the higher chloride concentration at the surface. However, the chloride content consistently decreased with increasing depth. They also found the penetrated chlorides, with 0.36% and 0.44% of cement content at 25 - 30 mm depth. However, chloride penetration can be minimized in OPS concrete using ultrafine SCM as a partial cement replacement.

It has been found that when subjected to actual marine conditions, the diffusion coefficient was found to be in the range of 7.1 - 32.5 × 10^{-8} cm^2/s and 6.6 - 10 ×10^{-8} cm^2/s for concretes with w/c ratio of 0.7 and 0.45 respectively [86]. The chloride diffusion coefficients of OPS concrete were in the range of 5.86 - 12.02×$10^{-8}$$cm^2/s$ [80]. These results show that the diffusivity of OPS concrete was comparable to that of normal weight concretes subjected to actual marine conditions.

5.11.7. Rapid Chloride Penetration Test

Teo et al. (2010) [80] conducted Rapid Chloride Penetration Tests (RCPT) using OPS concrete with 28, 56 and 90-day age samples in accordance with [87]. Their mix proportion was 1:1.66:0.60 (OPC:sand:OPS) by weight of 510kg/m^3 cement with w/c ratio of 0.38 including superplasticiser at rate of 1.4Litre/100kg cement. They used concrete samples of 100mm diameter and 50mm thick slices cut from middle portion of 100 mm diameter and 200 mm height concrete cylinder. They observed the 28-day RCPT values of OPS concrete ranged from about 3,581 to 4,549 Coulombs which showed moderate to high chloride-ion penetrability (Table 5.13). They also noticed increased temperatures after testing the samples. Stanish et al. [88], explained lower quality concrete in terms of compressive strength tended to heat more. So, the lower the quality of concrete, the greater was the current at given voltage resulting in higher heat gain. So, a further increase in the charge passed was due to heat gain in low strength concrete. Therefore, this may have been the reason for OPS concrete in this study [80].

Islam et al. (2016) [13], also studied RCPT using 90-day old OPS concrete and they found the value of 3,288 Coulombs. Their mix proportion was 1:1.7:0.65 (OPC:sand:OPS) by weight of 565kg/m^3 cement with w/c ratio of 0.3 including superplasticiser. They noticed the temperature development in NaCl cell from 29.4 ^{0}C (initial, at 1 minute) to 48^{0}C (final, at 360 minutes) and in NaOH cell from 29.2^{0}C (initial, at 1 minute) to 47.5^{0}C (final, at 360 minutes). In another study, the RCPT values of lightweight concrete made with expanded clay were in the range from 2,115-3,336 Coulombs [89].

Table 5.13. RCPT rating as per ASTM C1202 [87]

Charge passed (Coulombs)	Chloride ion penetrability
>4,000	High
2,000–4,000	Moderate
1,000–2,000	Low
100–1,000	Very low
<100	Negligible

5.11.8. Sulphate Resistance Capacity

The physical and chemical aspects of concrete are important when concrete is subjected to a sulphate environment [90]. Sulphate ions from external sources first diffuse into concrete and subsequently react with cement paste, leading to disintegration [91]. Sulphate salts of sodium, magnesium, ammonium or calcium exist naturally in soils, ground water and the sea at different concentrations. Sulphate solutions attack concrete. In a sulphate environment, concrete durability depends on many factors such as; type of structure, permeability of concrete, nature of ingress of sulphate, type and concentration of sulphate, water table and groundwater table [92]. Sulphate solution present in cement and in some aggregates in excessive amounts can cause expansion and disruption of concrete.

Sulphate solutions react with calcium hydroxide and calcium aluminate to form calcium sulphate (gypsum) and calcium sulfoaluminate (ettringite). Magnesium sulphate is found to be the most detrimental of all sulphates [93-95]. Magnesium sulphate disintegrates C-S-H gel to M-S-H gel, a non-cementitious and softening product. This reaction occurs in two steps [95]. The sulphate and magnesium react with C-S-H gel to form gypsum, brucite (magnesium hydroxide), silica gel and water in the first step of the reaction. In the second step, M-S-H gel and water are formed due to the reaction between brucite and silica gel. The mode of attack of sulphate on concrete is shown in Table 5.14 quantitatively [94, 96].

In water, sulphate concentration is very significant and controls the extent of damage on concrete structures. Previous works have used a wide range of sulphate solution of 1,000 ppm (0.1% concentration) to 72,000 ppm (7.2% concentration) [92-95, 97-100].

In line with [87], the magnesium sulphate solution with two concentrations namely 15,000ppm (1.5%) and 52,000ppm (5.2%) were used [2]. These two concentrations are in the very-severe category. The solution was changed at regular intervals in compliance with [87]. The 28-day age concrete cubes were immersed into this magnesium sulphate solution and tested. Tests were conducted at ages of 56 days and 90 days. Samples cured in W-1 and W-4 curing shown in Table 5.15 were employed for this study.

Table 5.14. Different sulphate concentrations in soils and water, and the type of attack on concrete [94, 96]

Type of attack	water-soluble sulphate (SO_4^{-2}) in soils (%)	sulphate (SO_4^{-2}) in water (ppm)
Negligible	0.00 to 0.10	0 to 150
Moderate	0.10 to 0.20	150 to 1,500
Severe	0.20 to 2.00	1,500 to 10,000
Very Severe	2.00 or more	10,000 or more

Table 5.15. Curing conditions for OPS concrete [2]

Curing symbol	Curing description	Duration of curing with place (day)		
		Mould	Water	Room ambient condition
W-1	Partial water	1	6	Rest days in room
W-4	Full water	1	Rest days in water	0

Figure 5.6 shows the compressive strength with age in tests of OPS concrete mixed with 0% FA, cured in W-1 and W-4 conditions [2]. This concrete was treated with 0ppm, 15,000ppm and 52,000ppm sulphate solution before compression tests. Here 0ppm sulphate solution indicates potable water without sulphate content. Comparisons were made with OPS concrete treated with 0ppm sulphate solution. OPS concrete cured in W-1 condition at its 90-day test showed the reduction in the strength of about 5.1% and 8.4% in 15,000ppm and 52,000ppm sulphate solutions, respectively. Similarly, for the same concrete cured in W-4 condition, the test result at 90-day, showed about 6.6% less strength under a 52,000ppm sulphate solution. Both comparisons were made with OPS concrete treated with 0ppm sulphate solution.

As shown in Figure 5.7, OPS concrete mixed with 10% FA, cured in W-1 condition, showed about 1.2% less strength at 90-day, under both 15,000ppm and 52,000ppm sulphate solution [2]. Similarly, the same concrete cured in W-4 condition, showed the strength reduction of 1.2% and 3.1%, for the 15,000ppm and 52,000ppm sulphate solution, respectively. The 52,000ppm sulphate solution showed more reduction in compressive strength than the 15,000ppm sulphate solution. OPS concrete mixed with 10% FA showed less strength reduction for both curing conditions.

The reduction in the compressive strength of OPS concrete was relatively high in 52,000ppm sulphate solution compared with 15,000ppm solution [2]. It indicated that strong sulphate solution disintegrated the paste. Sulphate solution can penetrate concrete more if there is a continuous flow inside concrete. OPS aggregate has a high water absorption capacity. This may bring the sulphate solution into OPS concrete. However, a detailed long-term study is required to add more information for the conclusive decision. In such a long-term investigation the durability properties of OPS aggregates and OPS concrete in severe environments like sulphate solutions should be carried out.

Islam et al. (2016) [13], also studied the sulphate attack using 56-day and 90-day old OPS concrete. Their mix proportion was 1:1.7:0.65 (OPC:sand:OPS) by weight of 565kg/m^3 cement with w/c ratio of 0.3 including superplasticiser. They used 5% (50,000ppm) $MgSO_4$ (magnesium sulphate) solution and a found reduction of compressive strength of 4.4% and 20.2% at age of 56-day and 90-day. They used cycles of 2 weeks immersion in such sulphate solution followed by 2 weeks drying in room conditions which caused a high percentage of strength reduction.

The compressive strength reduction of cement mortar in 21,000ppm sulphate solution was noticed about 62% up to the age of 1 year [100]. Soil with 1.0% sulphate concentration could not reduce the compressive strength of concrete, as investigated by [92], for up to 5 years.

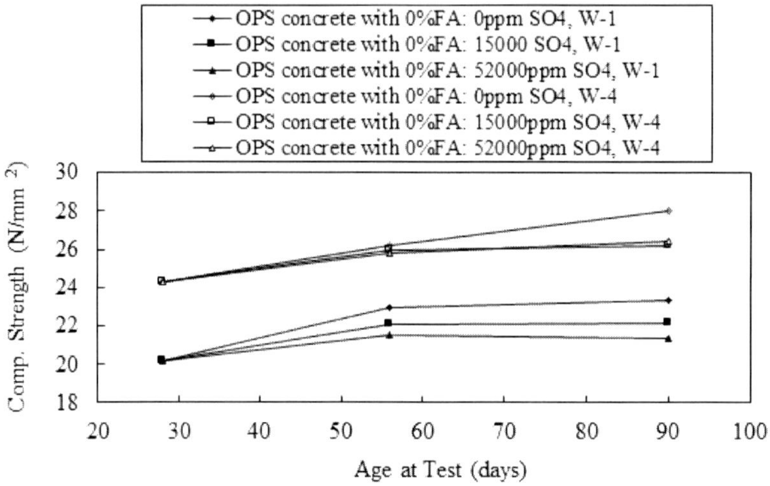

Figure 5.6. Compressive strength of OPS concrete mixed with 0% fly ash under different concentrations of sulphate solutions [2].

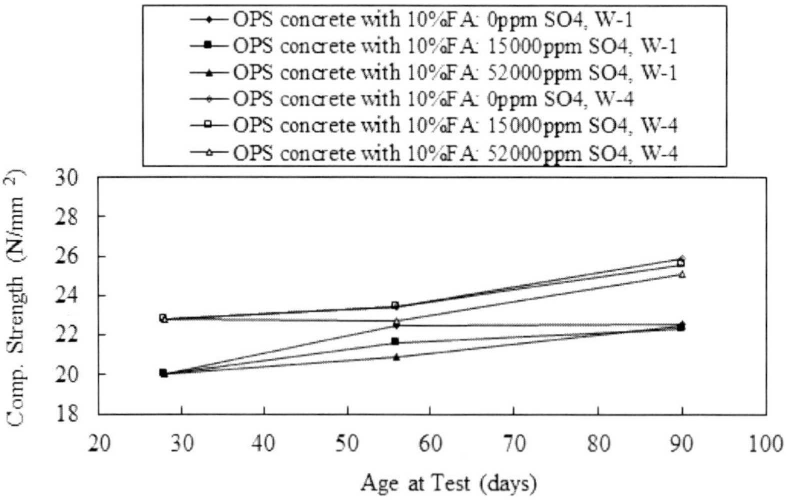

Figure 5.7. Compressive strength of OPS concrete mixed with 10% fly ash under different concentrations of sulphate solutions [2].

5.12. COST EFFICIENCY OF OPS CONCRETE

Figure 5.8 and Table 5.16 show the cost efficiency of OPS concrete based on findings [2, 3, 4, 9, 11, 35]. The unit cost of materials are obtained from [13, 101]. The highest cost efficiency of 0.109 N/mm²/RM/m³ is seen with [4] followed by 0.096 N/mm²/RM/m³ of [35], 0.096 N/mm²/RM/m³ of [11], 0.069 N/mm²/RM/m³ of [2], 0.061 N/mm²/RM/m³ of [3] and 0.054 N/mm²/RM/m³ of [9]. It should be noted that the cost of lightweight concrete made with artificial lightweight aggregate is always high.

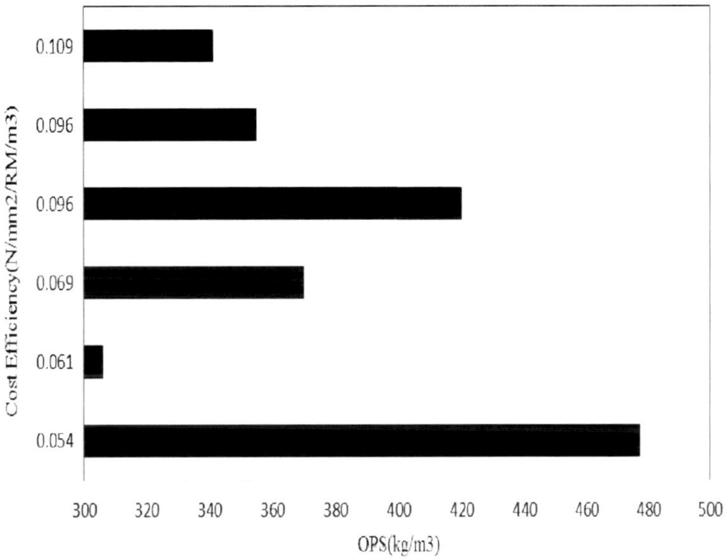

Figure 5.8. Cost efficiency OPS concrete (courtesy, author).

Table 5.16. Cost efficiency of OPS concrete [courtesy, author]

Reference, year and 28-day comp. St. (N/mm²)	OPC (kg/m3) [Cost =RM0.44/kg] (RM)	Sand(kg/m³) [Cost =RM0.08/kg] (RM)	OPS(kg/m³) [Cost =RM0.20/kg] (RM)	Normal stone(kg/m³) [Cost = RM0.06/kg] (RM)	SP(kg/m³) [Cost =RM15.28/kg (RM)	Total material cost/m³ (RM)	Cost efficien. [N/mm²/ RM/m³]
[2], 2001, 24.20	480 kg =RM211.20	820kg =RM65.60	370kg =RM74.00	-	-	350.80	0.069
[3], 2007, 28.00	510kg =RM224.40	847kg =RM67.76	306kg =RM61.20	-	7.14kg =RM109.10	462.46	0.061
[9], 2009, 26.98	596kg =RM262.24	596kg =RM47.68	477kg =RM95.40	-	5.96kg =RM91.07	496.39	0.054
[4], 2010, 38.53	450kg RM98.00	629kg =RM50.32	341kg =RM68.20	313kg =RM18.78	7.65kg =RM116.89	352.19	0.109
[11], 2013, 42.50	500kg =RM220.00	780kg =RM62.40	420kg =RM84.00	-	5kg =RM76.40	442.80	0.096
[35], 2016, 45.10	550kg =RM242.00	935kg =RM74.80	355kg =RM71.00	-	5.5kg =RM84.04	471.84	0.096

Notes: a) RM 4 = USD 1; b) Cost ref. for OPC, Sand, normal stone and SP [101]; c) Cost ref. for OPS [13]

5.13. Chapter Remarks

The investigations by many researchers have shown that OPS concrete is potential lightweight concrete for the construction industry. The detailed investigations have established the behaviour of OPS concrete. It indicates clearly under which situations OPS concrete behaves favourably and under what conditions it behaves unfavourably.

References

[1] M. L. Gambhir. 2004. *Concrete Technology*, 3rd edition. India. Tata McGraw-Hill Publishing Company limited.

[2] M. A. Mannan. 2001. *A study on the use of Oil Palm Shell in Lightweight Concrete*. PhD Thesis. Universiti Malaysia Sabah.

[3] D. C. L. Teo. 2007. *A study on structural lightweight concrete using Oil Palm Shell (OPS) aggregate*. PhD Thesis, Universiti Malaysia Sabah.

[4] C. H. Ng. 2010, *Studies on precast reinforced concrete floor panels using Oil Palm Shell aggregate*. PhD Thesis. Universiti Malaysia Sabah.

[5] R. J. Salter. 1979. *Highway Design and Construction*. London. The Macmillan Press Ltd.

[6] O. Fidelis O. 1988. Palm kernel shell as a lightweight aggregate for concrete. *Cement and Concrete Research*. **18.** pp. 901-910.

[7] D. C. Okpala. 1990. Palm kernel shell as a lightweight aggregate in concrete. *Building and Environment*. **25**. pp. 291-6.

[8] M. A. Mannan, J. Alexander, C. Ganapathy, D. C. L. Teo. 2006. Quality improvement of oil palm shell (OPS) as coarse aggregate in lightweight concrete, *Building and Environment* **41** pp. 1239–1242.

[9] H. Mahmud, M. Z Jumaat, U. J. Alengaram. 2009. Influence of Sand/Cement ratio on mechanical properties of palm kernel shell concrete. *Journal of Applied Sciences*. **9**. pp. 1764-1769.

[10] U. J. Alengaram, H. Mahmud, M. Z. Jumaat. 2011. Enhancement and prediction of modulus of elasticity of palm kernel shell concrete. *Materials and Design*. **32.** pp. 2143-2148.

[11] P. Shafigh, U. J. Alengaram, H. B. Mahmud, M. Z. Jumaat. 2013. Engineering properties of oil palm shell lightweight concrete containing fly ash. *Materials and Design*. **49.** pp. 617-621.

[12] J. Yahaghi, P. Shafigh, Z. C. Muda, S. B. Beddu. 2015. Influence of polypropylene fiber and hybrid fiber on mechanical properties of lightweight concrete. *Journal of Civil Engineering Research*. **5**. pp. 17-20.

[13] M. M. U. Islam, K. H. Mo, U. J. Alengaram. 2016. Mechanical and fresh properties of sustainable oil palm shell lightweight concrete incorporating palm oil fuel ash. *Journal of Cleaner Production*. **115**. pp. 307-314.

[14] J. L. Clarke. 1993. *Structural lightweight aggregate concrete*. Blackie Academic & Professional- an Imprint of Chapman & Hall.

[15] M. Sidney, J. F. Young. 1981. *Concrete*. Englewood Cliffs, New Jersey. Prentice-Hall Inc.
[16] B. H. Spratt. 1974. *The structural use of lightweight aggregate concrete.* Cement and Concrete Association.
[17] S. Andrew, W. Kinniburgh. 1978. *Lightweight Concrete*, 3rd edition. London. Applied Science Publishers Ltd.
[18] O. Fidelis O. 1991. An investigation on the use of superplasticizer in palm kernel shell aggregate concrete. *Cement and Concrete Research.* **21**. pp. 551-557.
[19] E. A. Olanipekun, K. O Olusola, O. Ata. 2006. A comparative study of concrete properties using coconut shell and palm kernel shell as coarse aggregates. *Building and Environment.* **41.** pp. 297-301.
[20] U. J. Alengaram, M. Z. Jumaat, H. Mahmud. 2008. Ductility behaviour of reinforced palm kernel shell concrete beams. *European Journal of Scientific Research.* **23**. pp.406-420.
[21] U. J. Alengaram, H. Mahmud, M. Z. Jumaat. 2010. Comparison of mechanical and bond properties of oil palm kernel shell concrete with normal weight concrete. *International Journal of the Physical Sciences.* **5**. pp. 1231-1239.
[22] FIP. 1983. *FIP Manual of Lightweight Aggregate Concrete*, 2nd edition. London. Surry University Press.
[23] D. F. Orchard 1979. *Concrete Technology*, all volumes, 4th edition. London. Applied Science Publishers Ltd.
[24] A. M. Neville. 1981. *Properties of Concrete*, 3rd edition. Pitman.
[25] BS 8110-2: 1997. *Structural Use of Concrete – Code of Practice for Special Circumstances*. London: British Standards Institution.
[26] ACI 213R-87, 1994. Guide for Structural Lightweight Aggregate Concrete. *ACI Manual of Concrete Practice*. Detroit: American Concrete Institute.
[27] EN 1992-1-1. 2004. Eurocode 2: Design of Concrete Structures – Part 1-1: General Rules and Rules for Buildings. *European Committee for Standardization.*
[28] A. M. Neville. 1999. *Properties of Concrete.* (4th edition). Malaysia: Longman.
[29] S. A. Salam. 1982. Lightweight concrete made from palm oil shell aggregates and rice husk. *Proceedings of A Regional Seminar.* Universiti Putra Malaysia. Sept. 15-17. 1982. pp. 177-196.
[30] A. A. A. Ali. 1984. Basic strength properties of lightweight concrete using agricultural wastes as aggregates, *International Conference on Low-Cost Housing for Developing Countries,* Roorkee, India. November, 1984.
[31] D. C. L. Teo, M. A. Mannan, V. J. Kurian. 2006. Influence of site curing on bond properties of reinforced lightweight concrete. *Journal of Civil Engineering Research and Practice*. **3.** pp. 9-19.
[32] K. H. Mo, U. J. Alengaram, M. Z. Jumaat. 2015. Utilization of ground granulated blast furnace slag as partial cement replacement in lightweight oil palm shell concrete. *Materials and Structures.* **48**. pp.2545-2556.
[33] A. Sharmin, U. J. Alengaram, M. Z. Jumaat, S. M. A. Kabir, I. I. Bashar. 2015. Engineering properties of oil palm shell and palm oil clinker based geopolymer concrete. *Rev. Téc. Ing. Univ. Zulia*. **38**. pp.34-48.
[34] M. A. Mannan, C. Ganapathy. 2002, Engineering properties of concrete with oil palm shell as coarse aggregate. *Construction and Building Materials.* **16**. pp 29-34.

[35] K. H. Mo, T. S. Chin, U. J. Alengaram, M. Z. Jumaat. 2016. Material and structural properties of waste-oil palm shell concrete incorporating ground granulated blast-furnace slag reinforced with low-volume steel fibres. *Journal of Cleaner Production*. **133**. pp. 414-426.

[36] C. Bob. 1992. How to get better curing. *Concrete*. **26**. pp. 22-25.

[37] S. Popovics. 1998. *Strength and Related Properties of Concrete-a quantitative approach*. Canada. John Wiley & Sons.

[38] D. L. Kantro, S. Brunauer, C. H. Weise. 1961. Development of surface in the hydration of calcium silicates. *Solid Surface and the Gas-Solid Interface*. American Chemical Society, Advance in Chemistry. **33**. pp 199-219.

[39] P. Klieger. 1958. Effect of mixing and curing temperature on concrete strength. *ACI Journal. Proc*. **54**. pp 1063-1081.

[40] S. Popovics. 1986. Effect of curing method and final moisture condition on compressive strength of concrete. *ACI Journal*. Proc. **83**. pp 650-657.

[41] P. Garces, E. G. Alcocel, S. Chinchon, C. G. Andreu, J. Alcaide. 1997. Effect of curing temperature in some hydration characteristics of calcium aluminate cement compared with those of Portland cement. *Cement and Concrete Research*. **27.** pp. 1343-1355.

[42] [H. J. Chen, T. Yen, T. P. Lia, Y. L. Huang. 1999. Determination of the dividing strength and its relation to the concrete strength in lightweight aggregate concrete. *Cement & Concrete Composities*. **21**. pp. 29-37.

[43] T. A. Holm. 1980. Physical Properties of High Strength Lightweight Aggregate Concretes. *Proceeding of the 2nd International Congress on Lightweight Conrcete*. London: Construction Press. pp.187 – 204.

[44] ACI 301. 2005. Specifications for Structural Concrete (ACI 301-05). *ACI Manual of Concrete Practice*. Detroit: American Concrete Institute.

[45] ASTM C 330. Standard specification for lightweight aggregates for structural concrete. *Annual Book of ASTM Standards*.

[46] K. S. Elliott. 2006. *Two-day Seminar and Workshop on Design and Construction of Precast Concrete Structures*, July 27 – 28, 2006, Kota Kinabalu, Sabah, Malaysia.

[47] Smorgon Steel Group, 2004, *Humeslab Technical Manual*, Australia.

[48] K. S. Elliott. 1996. *Multi-storey Precast Concrete Framed Structures*. London: Blackwell Science Ltd.

[49] K. J. Mun. 2007. Development and tests of lightweight aggregates using sewage sludge for nonstructural concrete. *Construction and Building Materials*. **21**. pp. 1583-1588.

[50] M. A. Aziz, S. D. Ramaswamy, C. K. Murthy. 1978. Lightweight concrete using cork granules. *Proceedings of the International Conference on Materials of Construction for Developing Countries*. Bangkok, Thailand. August 22-24, pp 181-189.

[51] M. S. Shetty. 1993. *Concrete Technology*. India. S. Chand and Company ltd.

[52] M. A. Mannan, K. Neglo. 2010. Mix design for oil-palm-boiler clinker (OPBC) concrete. *Journal of Science and Technology (Ghana)*. **30**. pp 111-118.

[53] C. H. Huang, S. Y. Wang. 2013. Application of water treatment sludge in the manufacturing of lightweight aggregate. *Construction and Building Materials*. **43**. pp. 174-183.

[54] A. Harold N. 1983. *Highway Materials, Soils and Concretes*, 2nd edition. Virginia. Reston Publishing Company.

[55] T. Ruenkrairergsa. 1987. Design of some concrete roads in Thailand. *Concrete Road Technology Course*. Malaysia. April 20-24.

[56] European Union – Brite EuRam III. 2000. *Mechanical Properties of Lightweight Aggregate Concrete*. Document BE96-3942/R23.

[57] B. J. Addis, M. G. Alexander. 1994. Cement-saturation and its effects on the compressive strength and stiffness of concrete. *Cement and Concrete Research*. **24**. pp. 975-986.

[58] M. G. Alexander, T. I. Milne. 1995. Influence of cement blend and aggregate type on stress-strain behaviour and elastic modulus of concrete. *ACI Materials Journal*. Technical paper. **92-M24**. pp. 227-235.

[59] M. G. Alexander. 1993. Two experimental techniques for studying the effects of the interfacial zone between cement paste and rock. *Cement and Concrete Research*. **23**. pp. 567-576.

[60] D. C. L Teo, M. A. Mannan, V. J. Kurian, 2005, Structural bond performance of lightweight concrete, *Proceedings Conference on sustainable building in South East Asia*, 11-13 April 2005, Kuala Lumpur, Malaysia.

[61] D. C. L. Teo, M. A. Mannan, V. J. Kurian, C. Ganapathy, 2007, Lightweight concrete made from oil palm shell (OPS): structural bond and durability properties, *Building and Environment,* **42**, pp 2614-2621.

[62] ASTM C 234, Standard test method for comparing concretes on the basis of bond developed with reinforcing steel, *Annual Book of ASTM Standards*.

[63] BS 8110. Structural use of concrete Part 1. Code of practice for design and construction. London: *British Standard Institution*.

[64] P. Paramasivam, Y. O. Loke. 1978. Study of sawdust concrete. *Proceedings of the International Conference on Materials of Construction for Developing Countries*, Bangkok, Thailand. August 22-24. pp. 169-179.

[65] C. O. Orangun. 1967. The bond resistance between steel and lightweightaggregate (Lytag) concrete. *Building Science*. **2.** pp. 21–8.

[66] Research Report. 2006. Performance of concrete made from agricultural and industrial solid wastes and by-products, IRPA research grant no. 03-02-10-0033-EA0031, Part-2, Lightweight concrete by D. C. L. Teo, Universiti Malaysia Sabah.

[67] J. Geiseler, H. Kollo, E. Lang. 1995. Influence of blast furnace cements on the durability of concrete structures. *ACI Material Journal*. Technical paper. **92-M27.** pp. 252-257.

[68] S. Nagataki. 1995. Concrete technology in Japan. *Concrete under severe conditions: Environment and loading*. E & FN Spon. **1** pp. 3-20.

[69] Z. Sawicz, S. S. Heng. 1996. Durability of concrete with addition of limestone powder. *Magazine of Concrete Research*. **48**. pp. 131-137.

[70] G. A. Khoury. 1992. Compressive strength of concrete at high temperatures: a reassessment. *Magazine of Concrete Research*. **44**. pp. 291-309.

[71] N. Jackson, R. K. Dhir. 1996. *Civil engineering materials*, 5th edition. London. Macmillan Press Ltd.

[72] M. Saad, S. A. Abo-El-Enein, G. B. Hanna, M. F. Kotkata. 1996. Effect of temperature on physical and mechanical properties of concrete containing silica fume. *Cement and Concrete Research*. **26**. pp. 669-675.

[73] A. M. Brandt. 1995. *Cement-based Composites: Materials, Mechanical Properties and Performance.* London. E & FN SPON.
[74] M. A. Mustafa. 1990. A study on the effect of various factors on drying shrinkage of concrete. *M. Sc. Thesis.* Dept. of Civil and Structural Engineering, Universiti Kebangsaan Malaysia.
[75] ASTM C 426. Standard test method for drying shrinkage of concrete block. *Annual Book of ASTM Standards.*
[76] P. C. Aitcin, A. Neville, P. Acker. 1997. Integrated view of shrinkage deformation. *Concrete International.* **19**, pp. 35-41.
[77] A. M. Neville.2008. *Properties of concrete.* 14th ed. Malaysia: CTP-VVP.
[78] J. M. Khatib, P. S. Mangat. 1995. Absorption characteristics of concrete as a function of location relative to casting position. *Cement and Concrete Research.* **25**. pp.999–1010.
[79] EuroLightCon. 2000. Properties of lightweight concretes containing Lytag and Liapor. *European Union—Brite Eu-Ram III,* Document BE96-3942/R8.
[80] D. C. L. Teo, M. A. Mannan, V. J. Kurian. 2010. Durability of lightweight OPS concrete under different curing conditions. *Materials and Structures.* **43**. pp.1–13.
[81] A. Shayan, A. Xu. 2006. Performance of glass powder as a pozzolanic material in concrete: a field trial on concrete slabs. Cement and Concrete Research. **36**. pp. 457–468.
[82] ASTM C 642. Standard test method for density, absorption, and voids on hardened concrete. *Annual Book of ASTM Standards.*
[83] H. Saricimen, M. Maslehuddin, A. J. Al-Tayyib, A. I. Al-Mana. 1995. Permeability and durability of plain and blended cement concretes cured in field and laboratory conditions. *ACI Structural Journal.* **92**. pp.1–6.
[84] Y. Maltais, E. Ouellet, J. Marchand, E. Samson, D. Burke. 2006. Prediction of the long-term durability of lightweight aggregate concrete mixtures under severe marine environment. *Material Structure.* **39**. pp. 911–918.
[85] AASHTO T 259. Standard method of test for resistance of concrete to chloride ion penetration. *American Association of State Highway and Transportation Officials.*
[86] S. K. Roy, K. C. Liam, D. O. Northwood. 1993. Chloride ingress in concrete as measured by field exposure tests in the atmospheric, tidal and submerged zones of a tropical marine environment. *Cement and Concrete Research.* **23**. pp. 1289–1306.
[87] ASTM C 1202. Standard test method for electrical indication of concrete's ability to resist chloride ion penetration. *Annual Book of ASTM Standards.*
[88] K. D. Stanish, R. D. Hooton, M. D. A. Thomas. 1997. Testing the chloride penetration resistance of concrete: a literature review, prediction of chloride penetration in concrete. *FHWA contract DTFH61-97-R-00022.* pp.1–30.
[89] K. S. Chia, M. H. Zhang. 2002. Water permeability and chloride penetrability of high-strength lightweight aggregate concrete. *Cement and Concrete Research.* **32**. pp.639–645.
[90] R. P. Khatri, V. Sirivivatnanon, J. L. Yang. 1997. Role of permeability in sulphate attack. *Cement and Concrete Research.* **27**. pp. 1179-1189.
[91] A. B. Yilmaz, B. Yazici, M. Erbil. 1997. The effects of sulphate ion on concrete and reinforced concrete. *Cement and Concrete Research.* **27**. pp. 1271-1279.
[92] E. F. Irassar, A. D. Maio, O. R. Batic. 1996. Sulphate attack on concrete with mineral admixtures. *Cement and Concrete Research.* **26**. pp. 113-123.

[93] T. Fikret, A. Fevziye, K. Sema, Y. Nabi. 1997. Effects of magnesium sulphate concentration on the sulphate resistance of mortars with and without silica fume. *Cement and Concrete Research*. **27**. pp. 205-214.

[94] A. Fevziye, T. Fikret, K. Sema, Y. Nabi. 1995. Effects of sodium sulphate concentration on the sulphate resistance of mortars with and without silica fume. *Cement and Concrete Research*. **25**. pp. 1360-1368.

[95] B. David, M. D. Cohen. 1992. Magnesium sulphate attack on Portland cement paste-i. microstructural analysis. *Cement and Concrete Research*. **22**. pp. 169-180.

[96] *Concrete Manual*. 1981. 8th edition. U.S. Department of the Interior Water and Power Resources Service. USA. John Wiley & Sons Publication.

[97] D. Mirjana, R. Jonjaua, O. Radovan, M. Sasa. 1996. Sulphate corrosion of Portland cement-pure and blended with 30% of fly ash. *Cement and Concrete Research*. **26**. pp. 1295-1300.

[98] B. David. 1993. A microstructural study of the effect produced by magnesium sulphate on plain and silica fume-bearing Portland cement mortars. *Cement and Concrete Research*. **23**. pp. 541-553.

[99] C. D. Lawrence. 1992. The influence of binder type on sulphate resistance. *Cement and Concrete Research*. **22**. pp. 1047-1058.

[100] A. O. S. Baghabra, M. Maslehuddin, M. M. Saadi. 1995. Effect of magnesium sulphate and sodium sulphate on the durability performance of plain and blended cements. *ACI Materials Journal*. Technical Paper. **92-M3**. pp. 15-24.

[101] B. H. Nagaratnam, M. E. Rahman, A. K. Mirasa, M. A. Mannan. 2016. Workability and heat of hydration of self-compacting concrete incorporating agro-industrial waste. *Journal of Cleaner Production*. **112**. pp.882-894.

Chapter 6

FLEXURAL PROPERTIES OF REINFORCED CONCRETE BEAM MADE WITH OPS AGGREGATE

Keywords: OPS concrete, lightweight concrete, reinforced concrete beam, flexural properties, instrumentation

6.1. INTRODUCTION

An experimental investigation has been conducted on prototype reinforced concrete beams made with OPS concrete. All beams having an effective span of 3,000mm with rectangular cross sections have a variable reinforcement ratio and effective depth. All beams have shown good ductility due to the high energy absorbing properties of OPS aggregate. It is found that the deflection is the major design limitation for OPS concrete beam. Cracks at service loads are all well within the maximum allowable limit for durability requirements. In general, beams with low reinforcement ratios have satisfied all the serviceability requirements. From the investigation, OPS concrete has shown promising results for use as structural members, especially for the construction of low cost houses in the oil palm plantation sector, rural housing etc. where OPS is found in abundance.

6.2. OIL PALM SHELL (OPS) - A POTENTIAL COARSE AGGREGATE FOR STRUCTURAL LIGHTWEIGHT CONCRETE

Lightweight concrete can be achieved through various methods such as (i) using a uniform size of coarse aggregates by omitting the fine aggregates entirely, (ii) introducing large voids to increase the porosity in concrete and (iii) incorporating lightweight aggregates. The use of lightweight concrete is primarily to reduce the overall dead load of a structure. With light weight, there will be less dead load for the structure to support. This will eventually result in smaller load bearing members and a corresponding reduction in the size of foundations. A considerable amount of cost savings for the foundation can be achieved especially where pilings are required or where poor soil conditions exist. Lightweight concrete also plays the crucial role in countries where the occurrence of earthquakes is unavoidable. By reducing the weight of the structure, the catastrophic seismic forces and inertia forces that influence the structures can also be eventually reduced as these forces are proportional to the weight of the structure.

Malaysia is currently the world leader in oil palm production with planted land of 5.64 million hectares [1] and generates over 5 million tons of waste oil palm shell (OPS) annually. Research efforts have been directed towards the potential use of OPS as coarse aggregates in lightweight concrete construction for the low income families especially at areas where OPS is in abundance, such as in oil palm plantations.

Generally, artificial lightweight concretes have higher production costs per unit volume as compared to normal weight concretes. However, with OPS concrete, there is about 17% cost reduction compared to normal weight concretes of the same grade [2]. This is attributed to the OPS aggregate which is readily available at no cost. At the age of 28-days, the developed compressive strength of OPS concrete as lightweight concrete is recorded with 28 N/mm^2 [3].

In the past, studies have been carried out on the structural performance of lightweight concrete beams [4-7]. However, these are limited to lightweight concretes with artificially produced inorganic lightweight aggregates. Studies on lightweight concrete with agricultural solid waste like OPS are still in their infancy, particularly on flexural property.

6.3. Mix Proportion of OPS Concrete

The acceptable mix proportion was selected based on the workability, strength, density, durability and cost requirements. The mix design for the OPS concrete in this prototype beam fabrication was based on an optimum mix proportion [2]. The acceptable mix comprised of 510 kg/m^3 OPC as binder, 848 kg/m^3 river sand as fine aggregate and 308 kg/m^3 OPS as coarse aggregate with a free water/cement ratio of 0.38. Due to the irregularity in the shape of OPS aggregates, the addition of superplasticiser was necessary to maintain the mix proportions at a given w/c ratio and to achieve a medium workability concrete. The amount of superplasticiser used was 1.4 litres per 100 kg cement and this was within the recommended dosage provided by the manufacturer.

6.4. Fabrication of 3.2m Long RC Beams

In this investigation, a total of nine RC beams comprising both single and double reinforced sections were fabricated and tested. The beams were categorized into three groups. The first group, designated with the letter *S* was single reinforced beams having the same effective depth of 200 mm with varying reinforcement ratios. The second group of beams comprised of single reinforced sections with varying reinforcement ratios and effective depths and these were designated with *SR*. The third group of beams were double reinforced sections, which were assigned with *D*. All test beams had rectangular cross sections with a total length of 3,200 mm and a span of 3,000 mm, with the reinforcement ratio and effective depth being the variables. Since OPS concrete is intended to be used in practical applications, it was necessary to ensure that the test beam dimensions were sufficiently large to simulate a real structural element. The beam sizes and lengths were also chosen to assure that the beams would fail in flexure.

The reinforcement details for each beam are illustrated in Table 6.1. When more than one layer of steel bars was required, a clear spacing of 20 mm was maintained between the layers. Larger diameter bars were used as the bottom layer when different sizes of bars were involved. The yield stress, f_y of bars are shown in Table 6.2. A clear cover of 25 mm was provided for all beams. For the single reinforced beams, the upper section was provided with minimal bars (nominal reinforcement) for constructive purposes. Sufficient stirrups were provided outside the loading zone to avoid failure in shear. The central zone between point loads had no stirrups to avoid confinement of the concrete and since this zone was, in theory, in pure bending, stirrups would not be necessary.

The formwork for the beams were constructed using wooden planks. Before the pouring of OPS concrete, the insides of the formwork were oiled to prevent the concrete from bonding to the sides during hardening. The OPS lightweight concrete was then placed and carefully compacted using a hand-held mechanical poker vibrator. For each group of beams, companion specimens which consisted of cubes, cylinders and beam prisms were tested on the same day as the beam test to determine the basic engineering properties of the beams. The compressive strength was determined using 100 mm cube specimens and 150 × 300 mm cylinders were prepared for determining the modulus of elasticity and splitting tensile strength. The beam prisms of 100 × 100 × 500 mm were used to obtain the modulus of rupture. The measured concrete strengths were based on an average value of three tested companion specimens.

All beams and companion specimens were cast under shade and cured under the same curing conditions. Immediately after casting, they were covered with polythene sheets for 20-28 hours to minimize the loss of moisture. After that, the formworks and the moulds were removed. The beams and companion specimens were kept moist and undercover for another 6 days with wet burlap. They were left at ambient laboratory conditions (Temp. of 22 to 28°C and RH of 74 to 88%) until the age of test.

Table 6.1. Details of prototype 3m span RC beams [8]

Beam No.	Effective depth, d (mm)	Beam size	Reinfor. in tension zone	Reinfor. in compression zone	Tension steel ratio, $\rho = A_s/bd$, %	Stirrup
S1	200	150 × 230	2T10	2R8	0.52	R6-150
S2	200	150 × 230	2T12	2R8	0.75	R6-150
S3	200	150 × 230	3T12	2R8	1.13	R6-150
SR1	225	150 × 255	2T10 + 1 T12	2R8	0.80	R6-150
SR2	250	150 × 280	3T12	2R8	0.90	R6-100
SR3	275	150 × 305	2T12 + 1T16	2R8	1.04	R6-80
D1	200	150 × 230	3T16	2T10	2.01	R6-100
D2	200	150 × 235	3T20	2T16 + 1T12	3.14	R8-100
D3	200	150 × 240	3T20 + 2T12	2T20 + 1T10	3.90	R8-80

Note: T- deformed steel bar
R- round steel bar.

Table 6.2. Yield stress of deformed steel bar [8]

Diameter of deformed steel bar, mm	f_y, N/mm^2
10	509
12	495
16	510
20	528

6.5. INSTRUMENTATION

Strain gauges (model: FLA-10-11) with 120 Ω resistance and a gauge factor of 2.0 were attached to the prepared surfaces of the steel bars at the midspan section of the beam to measure the strains (Figures 6.1 and 6.2). For determining the concrete strains, an external grid of Demec points and electrical strain gauges (model: PL-60-11) were attached to the concrete surface at the central region of the beams to measure the strains at different depths (Figures 6.3 and 6.4). The measurement area of the electrical strain gauges was limited to a length of 60 mm, while Demec gauges had 200 mm measurement length. On the other hand, strains values were automatically captured using the electrical strain gauges. The readings from the Demec gauges were recorded manually. Therefore, the use of both gauges served as a check for strain readings. The Demec gauge is shown in Figure 6.5. An additional strain gauge (model: PL-60-11) was placed at the top concrete surface of the beams at midspan. The 100 mm plunger travel LVDTs were positioned at several locations including one at midspan and two directly below the loading points to determine the vertical deformations. All of the electrical strain gauges and LVDTs were connected to a portable data logger after which the readings were captured by a computer at preset load intervals until failure of the beam occurred.

Figure 6.1. Surface preparation for attachment of steel strain gauges [8].

The end rotations of the beams were measured using a theodolite with an accuracy of 1 second (Figure 6.6). The theodolite was positioned on the beam exactly over the support point, which was held in place using a bolt drilled and epoxied into the beam. A measurement staff was placed at a predetermined distance from the beam to enable vertical readings to be taken. The crack was measured with a hand-held microscope with an optical magnification of ×40 and a sensitivity of 0.02 mm (Figure 6.7). The curvaturemeter was used for curvature ductility (Figures 6.8 and 6.9). The typical RC beam test set up is shown in Figure 6.10.

Figure 6.2. Typical attachment of electrical strain gauges of model: FLA-10-11 to steel rods [8].

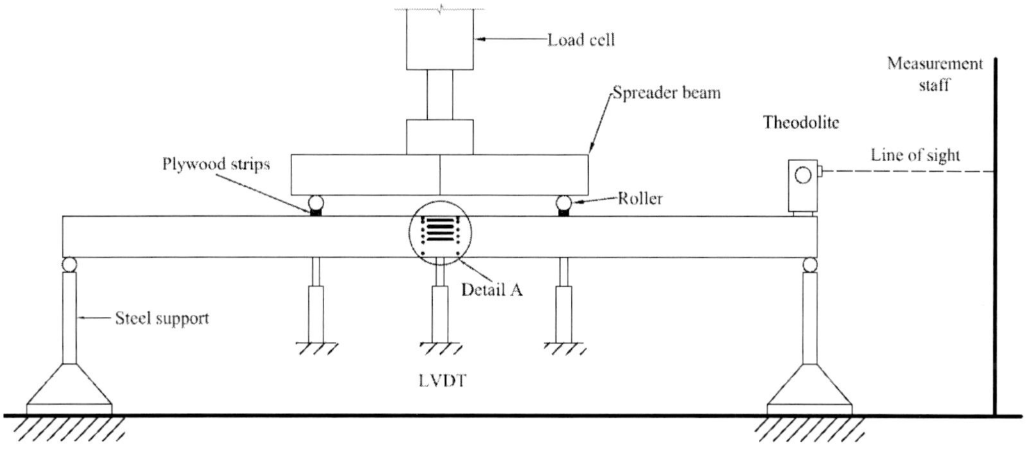

Figure 6.3. Beam test set-up [8].

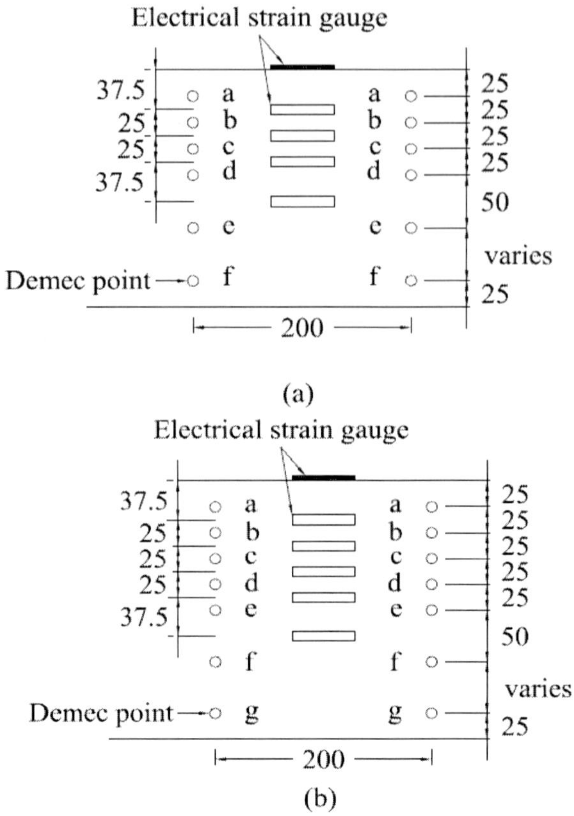

Figure 6.4. Detail A (Ref. to Figure 6.3) – Demec points and electrical strain gauge positions for beams (a) S1-S3, SR1 and D1-D3 (b) SR2 and SR3 [8].

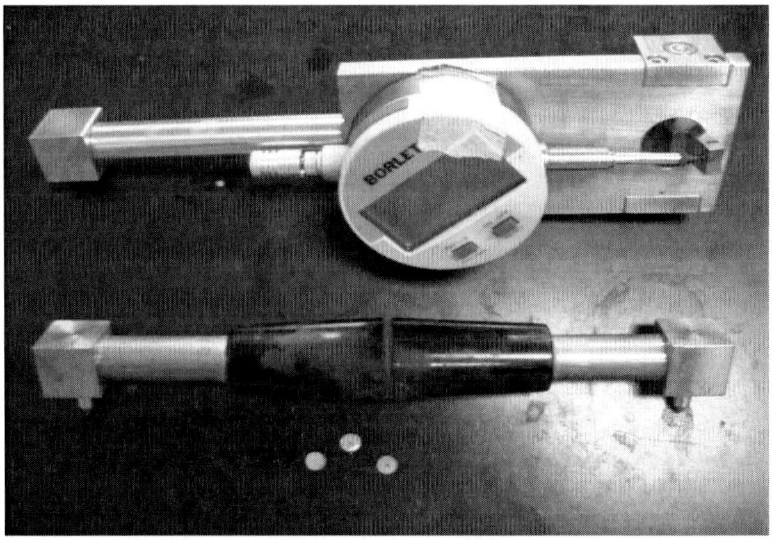

Figure 6.5. Demountable mechanical (DEMEC) strain gauge of digital read out with 200mm gauge length [8].

Figure 6.6. Theodolite of Nikon model with accuracy of 1 second [8].

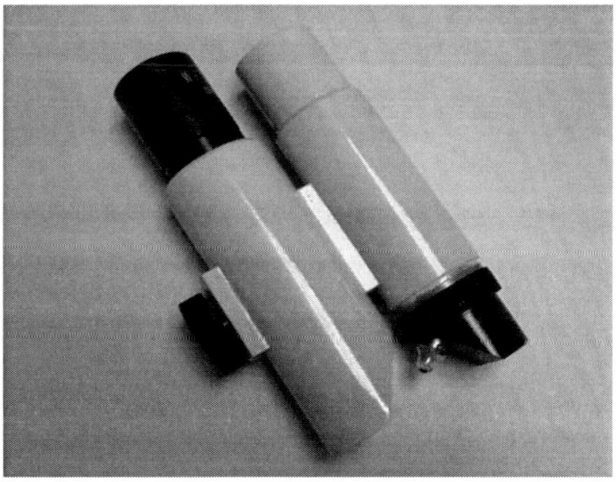

Figure 6.7. Hand-held microscope with an optical magnification of ×40 and a sensitivity of 0.02 mm [8].

Figure 6.8. Curvaturemeter, devised at UMS [8].

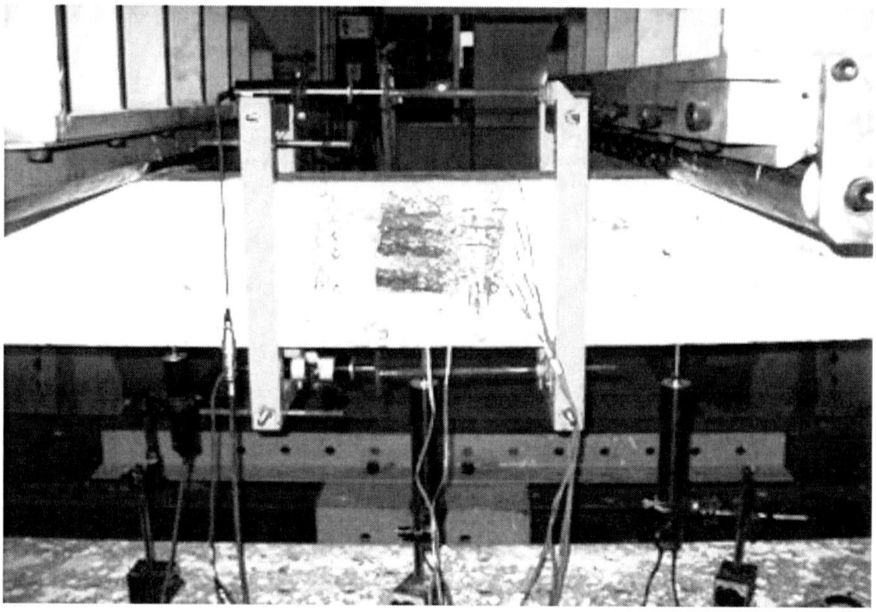

Figure 6.9. Typical set-up of curvaturemeter at mid span of RC beam [8].

Figure 6.10. Typical test of 3m span RC beam made with OPS lightweight concrete using Actuator of 1,000kN capacity. Author's lovely two PhD students, P. S. Kumar and D. C. L. Teo seen in investigation [8].

6.6. PROPERTIES OF OPS CONCRETE

The properties of OPS concrete conducted on the control specimens for each group of beams are shown in Table 6.3. The air-dry density of OPS lightweight concrete was below 2,000 kg/m^3 and the compressive strength was more than the minimum required for structural purposes [9]. This indicates that OPS concrete can be categorized as structural lightweight concrete.

OPS concrete has a low modulus of elasticity, which results in reduced stiffness as defined by the product of modulus of elasticity and moment of inertia, EI. Reduced stiffness can be beneficial for structures to a certain extent where differential foundation settlement may occur such as structures constructed on soft soils. In addition, the reduced stiffness is also advantageous for structures constructed in earthquake-prone areas.

Table 6.3. Properties of OPS lightweight concrete [8]

Beam series designation	S	SR	D
Beam type	Singly	Singly	Doubly
Air-dry density (kg/m^3)	1965	1955	1940
Compressive strength (N/mm^2)	26.3	25.0	25.3
Split tensile strength (N/mm^2)	1.82	1.51	1.67
Modulus of rupture (N/mm^2)	4.93	4.36	4.89
Static modulus of elasticity (GPa)	5.28	5.07	5.05

6.7. BEAM TEST SET-UP

The beams were simply supported and loaded in a Shimadzu, 1,000 kN capacity test machine until failure by two symmetrical concentrated loads. Figure 6.3 shows the schematic representation of the test set-up for a prototype RC beam. The rollers were adjusted so that the distance between the two load points was kept constant at 700 mm. Plywood strips were placed below the load points to absorb the irregularities of the top surface of the concrete beam. All beams were white-washed for checking the crack formation and propagation.

The beams were tested at an age of about 50 to 60 days. All test beams were subjected to monotonic loading under four-point bending. The loading was applied incrementally until the occurrence of failure was noticed. During testing, the beams were preloaded with a minimal load of 0.5 kN to allow initiation of the LVDTs and strain gauges. The beam was then unloaded and datum readings were recorded.

At each load increment the vertical deflection in the pure bending region, concrete strains and steel reinforcement strains were captured and the vertical displacement was obtained from the theodolite readings. The development of cracks was observed and the crack widths were measured using a hand-held microscope with an optical magnification of × 40 and a sensitivity of 0.02 mm.

6.7.1. General Behaviour of RC Beams

All beams exhibited typical failure in flexure with vertical flexural cracks appearing in the pure bending region. No horizontal cracks were observed at the level of the tension bars, which indicated that there was adequate bonding between the steel bars and the OPS concrete. Final failure was due to the gradual crushing of the compression concrete and the failures were not explosive. A global observation reveals that OPS concrete behave similarly to other lightweight aggregate concretes such as concrete with fly ash [6] and expanded slate [7].

Generally, the tensile reinforcement yielded before crushing of the concrete in the pure bending zone. Failure was gradual, with the beams undergoing a substantial amount of deflection prior to failure. The deflection at failure ranged between 61 mm to 78 mm. For any structural application, it is desirable that sections in flexure should be ductile and failure should occur with the gradual yielding of the tension steel and not by a sudden catastrophic failure of the concrete [10]. Once the reinforcement yields, plastic hinges are formed which allows the redistribution of bending moments to occur. As a consequence, more economical sections can be designed and safer structures can be achieved.

The summary of the test results for RC beams is presented in Table 6.4. In general, the beams exhibited a similar load-deflection pattern, with three distinct zones as illustrated in Figure 6.11. Before cracking occurred, the slope of the moment-deflection curve was steep and closely linear as indicated by segment 0-1. Here, the stiffness of the member is provided by the total concrete and steel area. However, once flexural cracks formed, there was a change in the slope of the load-deflection curve due to the reduction in stiffness (segment 1-2). This slope remained fairly linear until the yielding of the steel reinforcement took place. At the third stage (segment 2-3), the load-deflection curve demonstrated large increases in deflection with little increase in load. The experimental load-deflection curves are shown in Figures 6.12 to 6.14.

Table 6.4. Test beam results [8]

Beam No.	Cracking load, kN	Yielding load, kN	Ultimate load, kN	Deflection at failure, mm	Crack width at failure, mm	End rotation at failure
S1	4.00	20.00	28.00	72.18	1.24	3° 3' 9.79"
S2	5.00	27.50	38.50	73.40	0.90	3° 14' 1.68"
S3	6.50	45.00	58.00	77.20	0.82	3° 18' 42.48"
SR1	5.00	40.00	48.30	73.22	1.40	2° 5' 22.68"
SR2	9.00	60.00	64.50	67.55	1.80	2° 35' 2.47"
SR3	10.00	80.00	91.50	60.05	1.60	2° 38' 9.54"
D1	7.50	65.00	87.00	73.26	1.10	3° 20' 19.83"
D2	7.50	110.00	134.00	78.38	1.00	3° 6' 17.05"
D3	10.00	120.00	145.00	60.76	0.80	3° 6' 17.05"

It was observed that cracks extended quite rapidly towards the compression zone during the first few load increments but slowed down afterwards. For the single reinforced beams, initial cracking occurred at about 10 to 15% of the ultimate load, whereas for the double reinforced sections, the cracks formed at about 5 to 9% of the ultimate load. This reveals that for higher reinforcement ratios, the first crack occurs at a smaller percentage of the ultimate load. Lightweight concrete beams produced from fly ash coarse aggregate and expanded slate aggregate also exhibited similar trends [6, 7].

The end rotations of the beams were also determined and these reflect on the curvature of a beam. It can be observed that the end rotations at failure ranged from 2° 5' 23" to 3° 20' 19.83" and in general, these rotations are comparable to those obtained by [6].

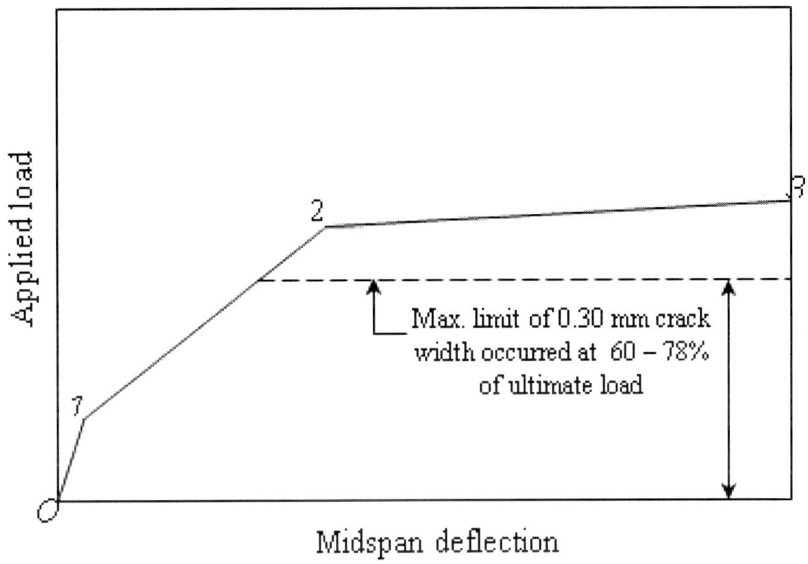

Figure 6.11. Typical load-deflection curve for RC beam made with OPS lightweight concrete [8].

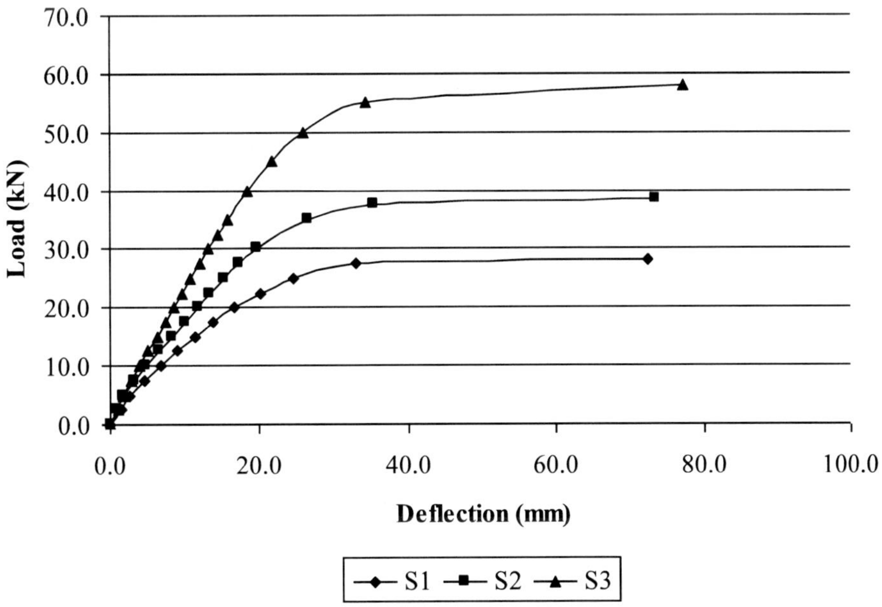

Figure 6.12. Load-deflection curve for S-series beams [8].

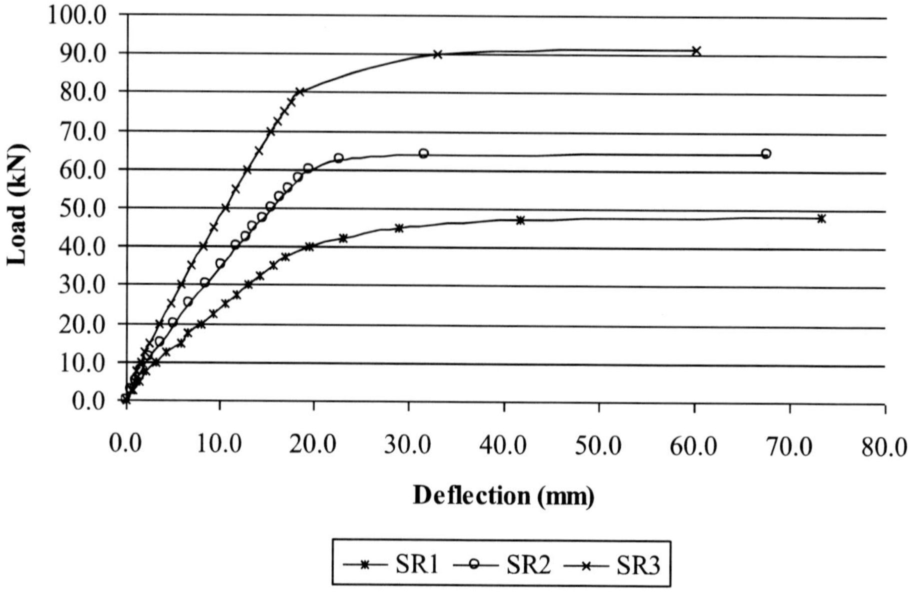

Figure 6.13. Load-deflection curve for SR-series beams [8].

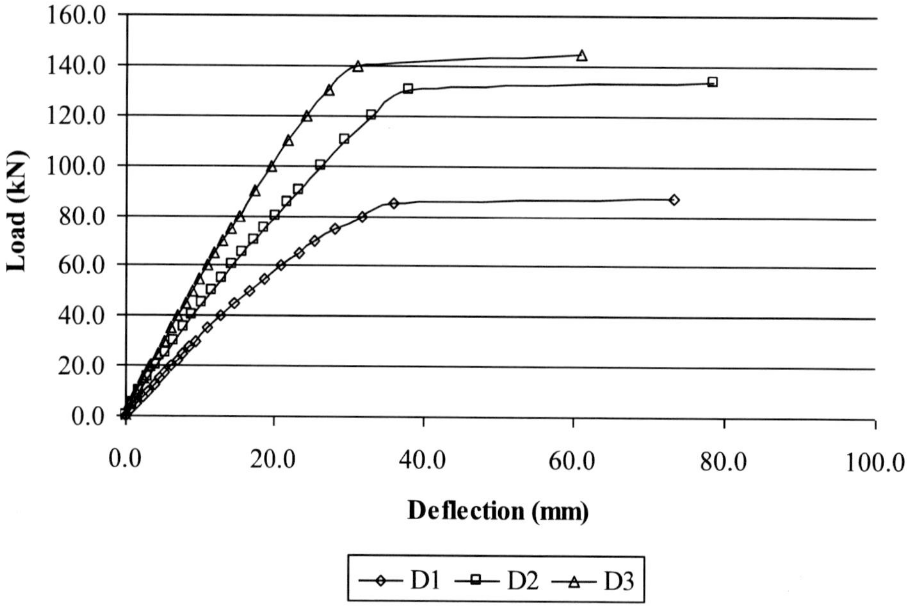

Figure 6.14. Load-deflection curve for D-series beams [8].

6.7.2. Crack and Deflection at Service Loads

For structural members to perform satisfactorily, there should not be excessive cracking and deflection at service loads. The wide cracks greatly reduce the stiffness of a member. It promotes ingress of harmful substances into the concrete and it also may cause concern to occupants. The excessive deflection may cause damage to other elements attached to the member. The crack widths and deflection at service load (obtained from the addition of beam self-weight and design ultimate load with a factor of 1.0) are presented in Tables 6.5 and 6.6 respectively. The crack widths and deflections were also compared to the theoretical values.

The maximum allowable crack widths provided by various codes of practice are usually in the range of 0.1 mm to 0.4 mm depending on the exposure condition. The [11] recommends that the surface crack widths should not be greater than 0.3 mm. At service loads, the crack widths of beams made with OPS concrete obtained from the investigation were all well below the maximum allowable crack width as recommended by [11]. The maximum allowable crack width of 0.3 mm generally occurred at about 60% to 78% of the experimental ultimate moment. The [12] and [13] codes were used for the prediction of crack widths. It was found that both codes gave reasonably close predictions of the crack width. However, [13] predicts the experimental crack widths of RC beams made with OPS concrete with better accuracy compared to [12].

As for the deflections, they were predicted using the beam curvatures according to [12], from the equation (6.1)

$$\Delta = K\ell^2\psi \tag{6.1}$$

where,

Δ = midspan deflection
K = a constant depending upon the distribution of bending moments of a member
ℓ = effective span and
ψ = curvature at mid span of beam

It was observed that the deflection obtained from the experiment at the service loads compares reasonably well to the predicted deflection recommended by [12]. Although OPS concrete has a low modulus of elasticity, the deflections at service loads for all single reinforced beams were within the maximum allowable value as stipulated by [11]. The [11] recommends an upper limit of span/250 for the deflection in order to satisfy the appearance and safety criteria of a structure. As expected, when the depth of the beams were increased (SR-series), the deflections were also reduced significantly.

6.7.3. Ultimate Strength

The experimental ultimate moment (M_{exp}) is compared with that predicted by the rectangular stress block analysis as per [11] (M_{pred}) and is presented in Table 6.7. It was observed that except for beam D3, all beams showed higher experimental ultimate moments compared to the predicted moments.

The values obtained from the current experiment were compared with the work of other investigators on lightweight concrete beams and these are shown in Table 6.8. It can be observed that the results obtained from the current investigation are consistent with the work of other researchers.

Table 6.5. Crack widths at service loads [8]

Beam No.	Service Load, kN	Experimental crack width (w_{exp}), mm	Theoretical crack width as per BS8110 (w_{theoBS}), mm	Theoretical crack width as per ACI ($w_{theoACI}$), mm	w_{exp}/w_{theoBS}	$w_{exp}/w_{theoACI}$
S1	15.05	0.22	0.19	0.23	1.16	0.96
S2	19.92	0.22	0.19	0.22	1.16	1.00
S3	27.15	0.22	0.19	0.22	1.16	1.00
SR1	26.50	0.23	0.13	0.21	1.77	1.10
SR2	35.47	0.18	0.13	0.18	1.38	1.00
SR3	47.89	0.17	0.16	0.21	1.06	0.81
D1	46.45	0.26	0.23	0.19	1.13	1.37
D2	80.56	0.26	0.30	0.23	0.87	1.13
D3	96.72	0.27	0.37	0.23	0.73	1.17

Table 6.6. Deflection at service loads [8]

Beam No.	Service Load, kN	Experimental deflection (Δ_{exp}), mm	Theoretical deflection (Δ_{theo}), mm	$\Delta_{exp}/\Delta_{theo}$	Span/Δ_{exp}
S1	15.05	11.40	10.75	1.06	263
S2	19.92	11.70	12.76	0.92	256
S3	27.15	11.90	13.90	0.86	252
SR1	26.50	11.10	11.05	1.00	270
SR2	35.47	10.45	11.15	0.94	287
SR3	47.89	10.05	8.52	1.18	299
D1	46.45	15.30	15.90	0.96	196
D2	80.56	20.50	18.35	1.12	146
D3	96.72	18.80	15.50	1.21	160

Table 6.7. Experimental and predicted moments [8]

Beam no.	Experimental ultimate moment, M_{exp} (kNm)	Theoretical design moment, M_{pred}, (kNm)	Ratio M_{exp}/M_{pred}
S1	16.10	13.60	1.23
S2	22.14	18.07	1.17
S3	33.35	24.73	1.35
SR1	27.77	24.10	1.15
SR2	37.09	32.32	1.15
SR3	52.61	43.72	1.20
D1	50.03	42.46	1.18
D2	77.05	73.87	1.04
D3	83.38	89.09	0.94

Table 6.8. Ultimate moment prediction for various lightweight concrete beams [8]

Investigators	Type of lightweight aggregate used	No. of beams tested	Code used for ultimate moment prediction	M_{exp}/M_{pred}	
				Mean	Standard deviation
Present study	Oil palm shell (OPS)	9	BS 8110	1.16	0.115
Lai et al., 1996 [4]	Polystyrene bead	1	AS 3600	1.16	-
Swamy and Lambert, 1984 [6]	Fly ash	6	CP 110	1.24	0.038
Swamy and Ibrahim, 1975 [7]	Expanded slate (Solite)	4	CP 110	1.59	0.166

6.7.4. Strain Distribution

Table 6.9 shows the summary of the strain characteristics of the tested OPS concrete beams. At every load increment, the concrete strain distribution over the depth of the beams was measured. It was found that strains obtained from both the DEMEC gauges and electrical strain gauges were closely similar. A typical strain distribution curve from each beam series is shown in Figures 6.15 to 6.17 in which the strains were obtained though DEMEC gauge readings. From the figures, it can be observed that the neutral axis depth progressively decreases as the loading applied increases. The figures also indicate that the strains were distributed comparatively linear above the neutral axis throughout the loading range. Once cracking was initiated in the concrete tension zone (below the neutral axis), the strain readings were discarded as this would not give a true representation of the strain values obtained. Before cracking, the concrete tensile strain obtained at 25 mm from the bottom tension face of the beams (DEMEC gauge 'F' or 'G') varied from 29 to 80 x 10^{-6}. In general, the results obtained are consistent with the works of other researchers [6, 7].

Table 6.9. Strain characteristics of RC beams made with OPS lightweight concrete [8]

Beam No.	Experimental Ultimate load, Lult (kN)	Effective depth, d (mm)	Neutral axis depth at experimental ultimate load, x (mm)	x/d	Tensile reinforcement strain			Max. recorded concrete compressive strain (x 10-6)
					Before cracking (x 10-6)	At yielding (x 10-6)	Max. recoded (x 10-6)	
S1	28.00	200	50.03	0.250	212	2146	5501	2105
S2	38.50	200	67.20	0.336	180	2509	6069	3263
S3	58.00	200	73.42	0.367	253	2932	6048	3605
SR1	48.30	225	81.46	0.362	220	2659	4145	2480
SR2	64.50	250	111.25	0.445	277	2574	3968	3638
SR3	91.50	275	121.38	0.441	270	2315	2839	2384
D1	87.00	200	137.01	0.685	204	1509	2705	5427
D2	134.00	200	140.18	0.701	111	2004	2643	5245
D3	145.00	200	155.30	0.777	170	1293	2093	5481

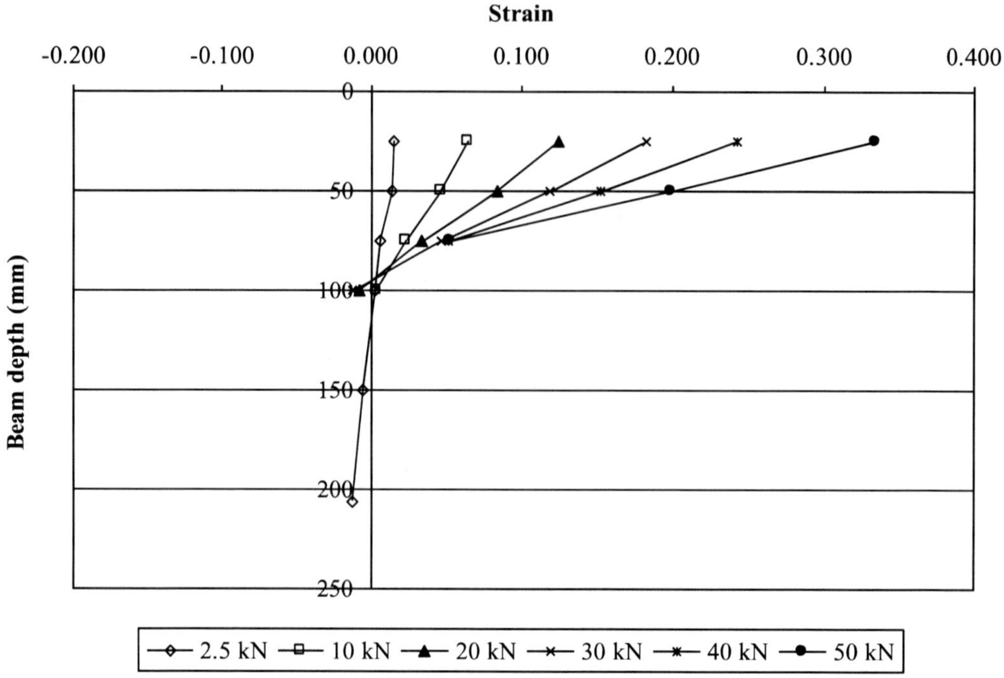

Figure 6.15. Strain distribution of beam S3 [8].

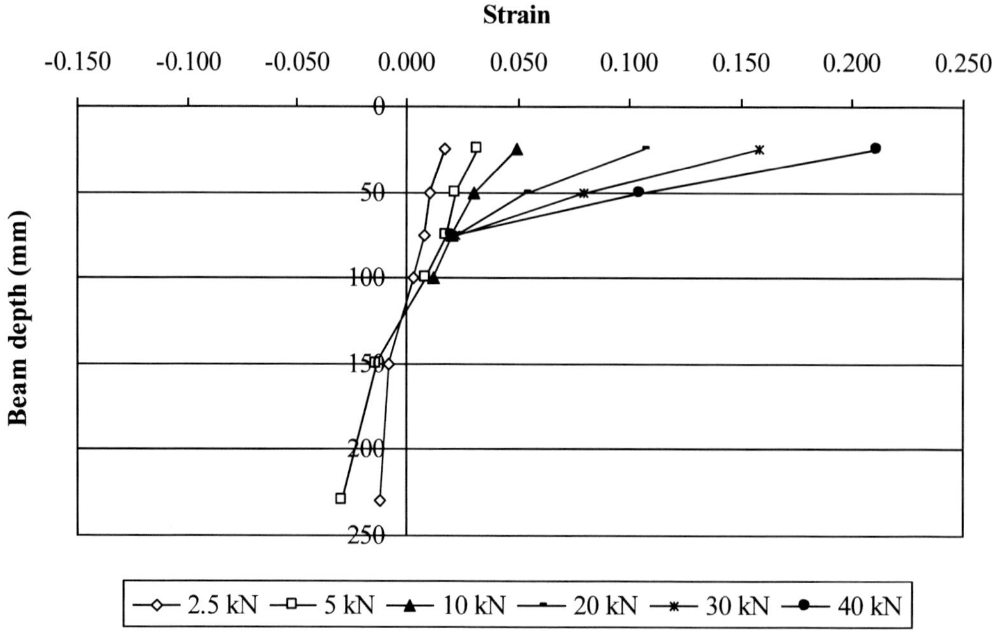

Figure 6.16. Strain distribution of beam SR1 [8].

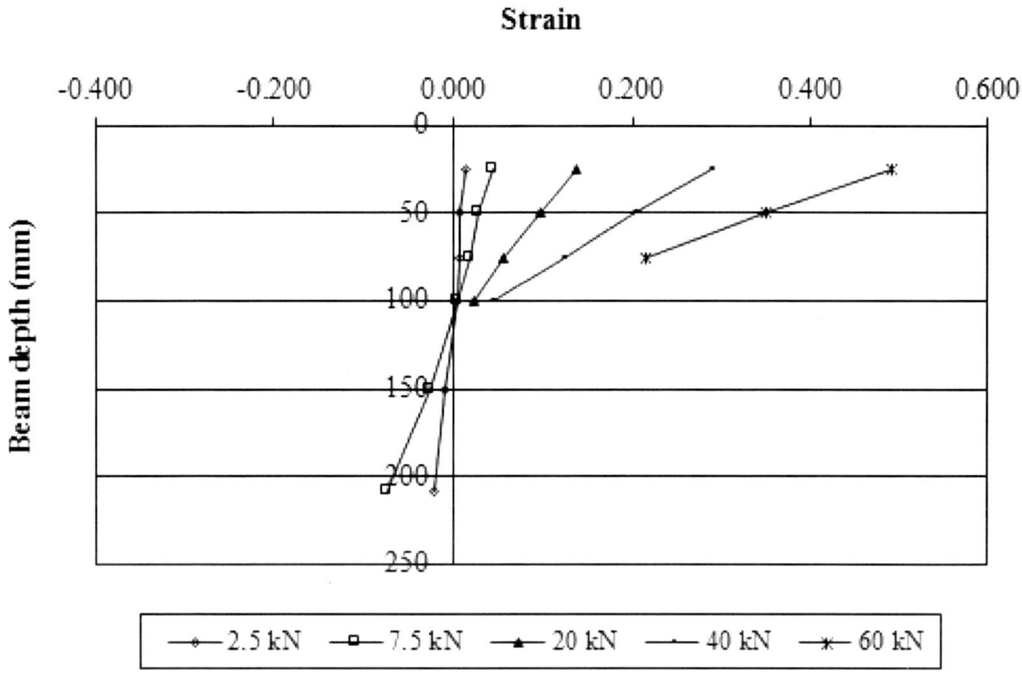

Figure 6.17. Strain distribution of beam D1 [8].

6.7.5. Ductility

Ductility may be defined as the ability to undergo deformations without a substantial reduction in the flexural capacity of the member [14]. Ductility is vital in structural members as it allows stress redistribution and provides a warning of impending failure. The redistribution of bending moments in the design of statically indeterminate structures enables more economical sections to be designed. The ductility of a beam can be measured in terms of displacement ductility, curvature ductility and rotational ductility. For this particular investigation, the ductility of these RC beams made with OPS concrete was determined through the curvature ductility index, $\mu=\psi_u/\psi_y$ which is the ratio of ultimate to first yield curvature. The curvatures of the beams were computed by dividing the extreme top fibre concrete compressive strains by the measured depth of the neutral axis. The neutral axis was obtained from the strains along the depth of each beam. Table 6.10 shows the ductility characteristics of RC beams made with OPS concrete. OPS concrete generally exhibits good ductility due to the nature of the OPS aggregates which have good energy absorbing qualities as indicated by the low AIV and ACV. A curvature ductility index of about 4 or more are typically considered for reinforced concrete members which are subjected to large displacements, such as forces caused by earthquakes [15].

Table 6.10. Curvature ductility characteristics of RC beams made with OPS lightweight concrete [8]

Beam No.	ψ_y (rad/m)	ψ_u (rad/m)	$\mu = \psi_u / \psi_y$
S1	0.0167	0.0729	4.36
S2	0.0204	0.0791	3.87
S3	0.0262	0.0957	3.66
SR1	0.0199	0.0625	3.14
SR2	0.0210	0.0640	3.05
SR3	0.0157	0.0452	2.87
D1	0.0247	0.0753	3.05
D2	0.0279	0.0713	2.55
D3	0.0217	0.0499	2.31

6.8. Chapter Remarks

The investigation on prototype reinforced concrete beams using OPS concrete as lightweight concrete has shown the promising structural behaviour which are useful to verify the applicability of current design provisions.

References

[1] Malaysian Palm Oil Board (MPOB); 2015. Oil palm planted area 2015. http://bepi.mpob.gov.my/images/area/2015/Area_summary.pdf: Retrieved on 14 July 2016.

[2] Teo DCL, Mannan MA, Kurian VJ. Influence of site curing on bond properties of reinforced lightweight concrete, *Journal of Civil Engineering Research and Practice* 2006; 3(1): 9-19.

[3] Teo DCL, Mannan MA, Kurian VJ. Structural Performance of Lightweight Concrete. *Proceedings of the Conference on Sustainable Building Southeast Asia. Kuala Lumpur, Malaysia*, 11-13 April 2005: 254-258.

[4] Lai KL, Ravindrarajah RS, Pasalich W, Hall B. Deformation behaviour of reinforced polystyrene concrete beam. *Proceedings of the Second International Conference on Advances in Composites.* Bangalore, India, 1996.

[5] Bungey JH, Madandoust R. Strength variations in lightweight concrete beams. *Cement and Concrete Composites* 1994; 16(1): 49-55.

[6] Swamy RN, Lambert GH. Flexural behaviour of reinforced concrete beams made with fly ash coarse aggregates. *The International Journal of Cement Composites and Lightweight Concrete* 1984; 6(3): 189-200.

[7] Swamy RN, Ibrahim AB. Flexural behaviour of reinforced and prestressed Solite structural lightweight concrete beams. *Building Science* 1975; 10(1): 43-56.

[8] Research Report. 2006. Performance of concrete made from agricultural and industrial solid wastes and by-products, IRPA research grant no. 03-02-10-0033-EA0031, Part-2, Lightweight concrete by D. C. L. Teo, Universiti Malaysia Sabah.
[9] ASTM C 330. Standard specification for lightweight aggregates for structural concrete. Annual Book of ASTM Standards.
[10] Mosely WH, Bungey JH. Reinforced concrete design. 4th Edition. Macmillan Press Limited; 1990.
[11] BS 8110. Structural use of concrete Part 1. Code of practice for design and construction. British Standard Institution. London.
[12] BS 8110. Structural use of concrete Part 2. Code of practice for special circumstances. British Standard Institution. London.
[13] ACI 318. Building code requirements for reinforced concrete. Detroit: American Concrete Institute.
[14] Park R, Ruitong D. Ductility of doubly reinforced concrete beam sections. *ACI Structural Journal* 1988; 85(2): 217-225.
[15] Leet K, Bernal D. Reinforced concrete design. 3rd Edition. McGraw-Hill; 1997.

Chapter 7

FLEXURAL BEHAVIOUR OF PRECAST RC FLOOR PANEL MADE WITH OPS LIGHTWEIGHT AGGREGATE

Keywords: Oil Palm Shell (OPS), OPS hybrid lightweight concrete (OPS hybrid LWC), OPS lightweight concrete (OPS LWC), Precast RC floor panel, solid panel, semi-precast panel, C-channel, lifting and transportation, flexural properties

7.1. INTRODUCTION

Precast RC solid floor panels of 3m span were investigated. Two types of lightweight concrete using 70% OPS and 30% granite, and 100% OPS as coarse aggregate were used. Both concretes being lightweight were named as OPS hybrid lightweight (OPS hybrid LWC) concrete and OPS lightweight concrete (OPS LWC). Two types of steel reinforcement used were namely (i) high-yield and (ii) welded-wire mesh. The investigated parameters included were panel weight, service and ultimate load capacity, deflection, crack spacing and width, ductility, concrete and steel strain characteristics. Results showed that both OPS hybrid LWC and OPS LWC panels reinforced with welded-wire mesh failed in a brittle manner. OPS hybrid LWC panel achieved a higher ultimate and service load capacity than OPS LWC panel. Also, solid panels made with OPS LWC showed more deflection than OPS hybrid LWC.

The precast RC solid panels of 3m span made with lightweight concrete were investigated for lifting and transportation. The panels were made using OPS hybrid LWC and OPS LWC. They were reinforced with high yield steel. For each floor panel, four Deha spherical-head plate lifting anchors were installed for lifting purpose. They were tested for lifting and transportation at an age of 7 days. The surface crack, compressive strength and strain distribution were determined. The solid panels made with OPS hybrid LWC satisfied the strength requirements at the time of lifting/transportation and at servicing. The impact of lifting and transporting was found to be insignificant.

The semi-precast RC floor panels of 3m span were investigated. Both OPS hybrid LWC and OPS LWC were used. The RC floor panel consisted of two layers, each 50 mm thick concrete where the topping concrete was made with either lower or the same concrete grade compared to that of the base concrete layer. When the base concrete reached the age of 7 to 18 days, topping concrete was cast upon it to form composite slab. The welded wire reinforcement was used. No shear connector was employed. Test parameters included were crack, deflection, moment and service load capacity. Results were also compared to RC floor panels made with normal weight concrete (NWC). No bond failure was observed between OPS aggregate and reinforcements. All reinforced concrete floor panels failed in brittle manner except the panel completely made with OPS LWC. Comparing to NWC, the average crack spacing of concrete base made with OPS LWC or OPS hybrid LWC was smaller under a higher ultimate load. The maximum no. of cracks between loading lines for OPS LWC or OPS hybrid LWC base was generally higher than NWC base due to a lower modulus of rupture for OPS LWC than NWC. The RC floor panels made with OPS LWC or OPS hybrid LWC base showed about 29% higher moment capacity resulting from better bonding with the topping concrete compared to that of NWC base. Regardless of the type of concrete base and topping, all floor panels marginally fulfilled the serviceability limit for residential building.

The new type of lightweight precast RC floor panel referred as 'C-channel' of 3 to 8m spans was investigated using lightweight concrete (LWC). The oil palm shell (OPS) was used as coarse aggregate in LWC. The OPS LWC of G28 and OPS hybrid LWC of G22 to G38 were used. OPS hybrid LWC made with 70% OPS and 30% granite by volume as coarse aggregate. The investigations were focused on concrete usage, weight and flexural properties of both ultimate and service conditions. It was observed that 'C-channels' were more cost effective and efficient in terms of saving the concrete, reduction of weight, higher load carrying capacity as compared with conventional full precast and semi-precast reinforced floor panels. The longer the span, the higher the concrete savings.

7.2. Precast Solid Floor Panel Reinforced with High-Yield Steel and Welded-Wire Mesh

An experimental investigation was conducted on RC floor panels of 3m span made with lightweight concrete using hybrid coarse aggregate with 70% OPS and 30% granite by volume, and with 100% OPS by volume. Both concretes being lightweight were named as OPS hybrid LWC and OPS LWC. Two types of steel reinforcement used were (i) high-yield and (ii) welded-wire mesh. It was to investigate the comparative performances of both OPS hybrid LWC and OPS LWC panels considering panel weight, service and ultimate load capacity, deflection, crack spacing and width, ductility, concrete and steel strain characteristics.

7.2.1. Materials and Experimental Investigation

The physical and mechanical properties used are shown in Table 7.1. Malaysian ordinary Portland cement (ASTM Type I) was used as binder, sand as fine aggregate and, OPS and crushed granite as coarse aggregates. The mix proportions are shown in Table 7.2. The shell thickness of OPS aggregate was 3.1 to 3.5mm. OPS aggregate was sieved and only aggregate passing through the 12.5 mm sieve and retained on the 4.75 mm sieve was used in concrete mixing. The w/c ratio was 0.38. Darex Super 20, a superplasticizer complied with specification [1], was incorporated in the mix to increase the fresh properties. OPS aggregate used in this investigation was in the saturated surface dry (SSD) condition. The tests of aggregates were done as per relevant BS and ASTM Standards. Tables 7.3 to 7.5 show the properties of OPS hybrid and OPS lightweight concretes.

Table 7.1. Physical and mechanical properties of Coarse and Fine aggregates [2]

No	Property of Aggregates	Coarse Aggregate		Fine Aggregate
		OPS	Granite	River Sand
Physical properties				
1	Maximum size, mm	12.50	19.00	1.18
2	Shell thickness	3.1 -3.5	-	-
3	Specific gravity (ssd condition)	1.34	2.63	2.43
4	Water absorption capacity, %	21.30	2.04	3.89
5	Fineness Modulus	6.30	6.94	1.33
Mechanical properties				
6	Aggregate Impact Value, AIV, %	6.60	12.60	~
7	Aggregate Crushing Value, ACV, %	7.12	15.95	~
8	Los Angeles Abrasion Value, %	2.90	11.85	~

Table 7.2. Mix ratio by weight of cement for 1.0 m^3 concrete (OPC =450kg/m^3) [2]

Ingredients	OPS hybrid lightweight concrete (OPS hybrid LWC)	OPS lightweight concrete (OPS LWC)
OPC (ASTM Type I)	1	1
w/c ratio	0.38	0.38
River sand as fine aggregate	1.40	1.40
OPS as coarse aggregate	0.76	1.08
Granite as coarse aggregate	0.70	-
Sp (%)	1.67	1.67

Table 7.3. Properties of different lightweight concretes at 28-day [2]

Properties	Type of concrete	
	OPS hybrid LWC	OPS LWC
Air-dry density (kg/m^3)	2,000	1,850
Unit weight (kN/m^3)	20.0	18.5
Compressive strength (MPa)	38.50	27.60
Split tensile strength (MPa)	3.50	1.70
Flexural strength (MPa)	5.80	3.80
Static modulus of elasticity, E (MPa)	21,100 (21.1GPa)	16,370 (16.37GPa)

Table 7.4. Compressive strength development of OPS hybrid LWC and OPS LWC with age [2]

Type of concrete	Development of compressive strength with age (MPa)				
	1-day	3-day	7-day	28-day	365-day
OPS hybrid LWC	17.30	25.80	35.97	38.50	39.50
OPS LWC	13.77	22.10	24.35	27.60	28.60

Table 7.5. Comparative strength development based on 28-day compressive strength [2]

Type of concrete	Percentage increase of compressive strength based on 28-day strength (%)				
	1-day	3-day	7-day	28-day	365-day
OPS hybrid LWC	44.94	67.01	93.43	100	102.60
OPS LWC	49.89	80.07	88.22	100	103.62

7.2.2. Panel Fabrication, Instrumentation and Test

Six solid RC floor panels with dimensions shown in Table 7.6 were cast and cured for 7 days under the shed outside the laboratory (ambient temp. of 28 ± 5°C, relative humidity of 68 to 91%). They were designed as one way panels. The floor panels were either reinforced with high yield steel bars 4T10 at equal spacing as main bar and T10 at 300 mm spacing for secondary reinforcement, or welded wire mesh type A10 with 200 mm spacing. In practice, welded wire mesh has been increasingly used for slabs because of the ease of placing the fabric sheets, control of reinforcement spacing, and better bonding. The yield strength of the high yield steel reinforcement and the welded wire mesh was approximately 460 and 485 N/mm^2 respectively. The steel ratio ($\rho = {100A_s}/{bd}$) ranged between 0.17 and 0.36%. The concrete cover was maintained at 20 mm.

The companion cubes of 100mm size were cast for each panel [3]. Prior to testing at 28-days, the sides of the floor panels were painted with white colour for clear detection of cracks. The crack width was measured after each applied load using a hand held microscope with an optical magnification of ×40. The panels were tested as simply supported under 2-line loading (Figure 7.1) at load increments of 2 kN up to 10 kN, and then 5 kN until failure using an actuator of 1,000 kN capacity. The distance between the loading points was kept constant at 1,000 mm. The Linear Variable Displacement Transducers (LVDTs) capable of measuring maximum displacements of 100 mm were used for deflection at midspan of the panel.

The deflection can be predicted using the BS 8110 [4]. The BS predicts deflection based on the Eq. (7.1):

$$\Delta = K\ell^2 \frac{1}{r_b} \tag{7.1}$$

where

Δ = midspan deflection

K = a constant that depends on the shape of the bending moment diagram

ℓ = effective span of panel

$\frac{1}{r_b}$ = curvature at midspan [$= \frac{M}{E_c I}$ where M is moment at the section considered]

E_c = modulus of elasticity of concrete, and

I = second moment of area of the gross uncracked section

On the other hand, the prediction of maximum deflection based on ACI Code which considers the cracked section can be used as in the Eq. (7.2):

$$\Delta_{max} = \frac{23PL^3}{648E_c I_e} \tag{7.2}$$

where,

P = applied load (two concentrated loads at one-third points)

L = clear span, and

I_e = effective moment of inertia shown in Eq. (7.3):

$$I_e = \left(\frac{M_{cr}}{M_a}\right)^3 I_g + \left[1 - \left(\frac{M_{cr}}{M_a}\right)^3\right] I_{cr} \le I_g \tag{7.3}$$

where,

M_{cr} = cracking moment (i.e. $\frac{f_r I_g}{Y_t}$)

f_r = modulus of rupture of concrete

M_a = maximum unfactored moment

I_g = moment of inertia of gross concrete section about the centriodal axis (neglecting the reinforcement)

I_{cr} = moment of inertia of cracked transformed section, and

Y_t = distance from centroidal axis of cross-section (neglecting steel) to tension face

The moment that produces a tensile stress just equal to the modulus of rupture is termed as the cracking moment M_{cr}. For a rectangular section of width b and a total depth h with tension steel A_s, I_r can be computed from Eq. (7.4):

$$I_{cr} = \frac{bx^3}{3} + nA_s(d-x)^2 \tag{7.4}$$

where,

x = distance of the neutral axis from the extreme compression fibers (x = $\frac{bx^2}{2} - nA_s(d-x)^2$

d = effective depth, and

n = modular ratio i.e., the ratio of elastic modulus of steel to concrete

For panels S1, S2, S4 and S5, four TML electrical strain gauges (Model: PL-60-11 with electric resistance of 120 Ω) SG2, SG3, SG4 and SG5 were fixed on the concrete's side at a depth of 25, 45, 65 and 85 mm respectively (Figure 7.2). Five similar strain gauges were used in panels S3 and S6. One electrical strain gauge (Model: FLA-10-11 with electric resistance of 120 Ω) SG6 was attached on the prepared surface of main steel bar at the midspan of panel. The LVDTs and strain gauges were connected to a portable data logger for recording the readings at preset load intervals until failure of the panel occurred. The general behaviour of the panels was carefully observed during the entire loading process.

At each load interval, the crack and crack propagation were marked on the side of the panels. According to [5], the maximum crack width at the level of the reinforcement where corrosion is expected can be predicted using the Eq. (7.5):

$$w_{max} = \frac{1.1 \times \sqrt[3]{t_b A}}{1 + \frac{2}{3}(t_s / h_1)} f_s \times 10^{-5} \tag{7.5}$$

where,

t_b = distance from extreme tension fiber to center of adjacent bar (mm)

t_s = distance from side of beam to center of adjacent bar (mm)

A = average effective area of concrete in tension around each reinforcing bar (A_e/n, where A_e is the rectangular concrete area of width b and with same centroid as steel, b the beam width, and n the number of bars)

f_s = steel stress (N/mm^2)

h_1 = distance from centroid of steel to neutral axis (mm)

The maximum probable surface crack-width can be determined using Eq. (7.6)

$$w_{cr} = 11 \times 10^{-6} \sqrt[3]{t_s \left(\frac{A_e}{n}\right)} \left(\frac{D-x}{d-x}\right) f_s \tag{7.6}$$

where,

D and d = total and effective depth of slab respectively, and

x = neutral axis of the section.

The end rotation was measured using a theodolite (Model NT-4D) with an accuracy of 1 second. The theodolite was positioned on the panel over the support point (Figure 7.1) and a staff was placed some distance from the theodolite to observe vertical readings at every load increment. It was used to determine the ductility of the panels.

Table 7.6. Solid RC floor panels [2]

Panel mark	Concrete	Panel dimension			Steel type and ratio	
		Width (mm)	Length (mm)	Thickness (mm)	Steel type	Steel ratio, ρ (%)
S1	OPS hybrid LWC	1,000	3,100	135	A10	0.36
S2					T10	0.24
S3				175	T10	0.17
S4	OPS LWC			135	A10	0.36
S5					T10	0.24
S6				175	T10	0.17

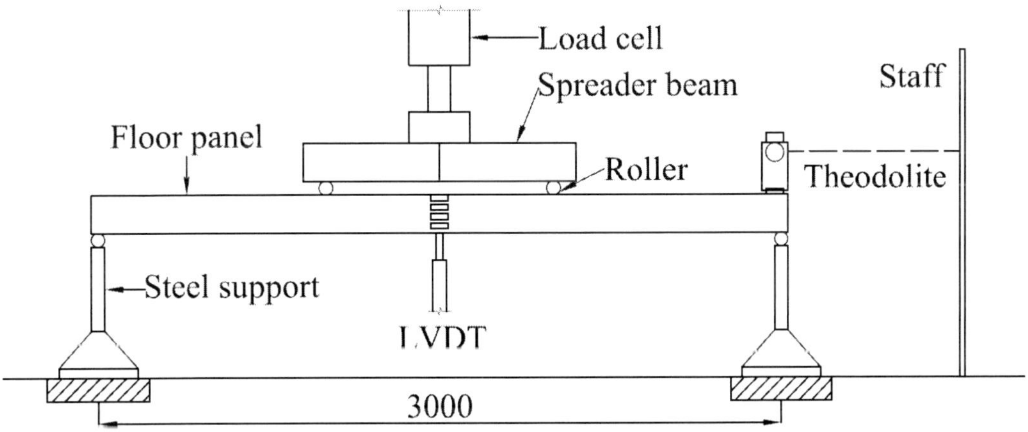

Figure 7.1. Schematic floor panel test setup [2].

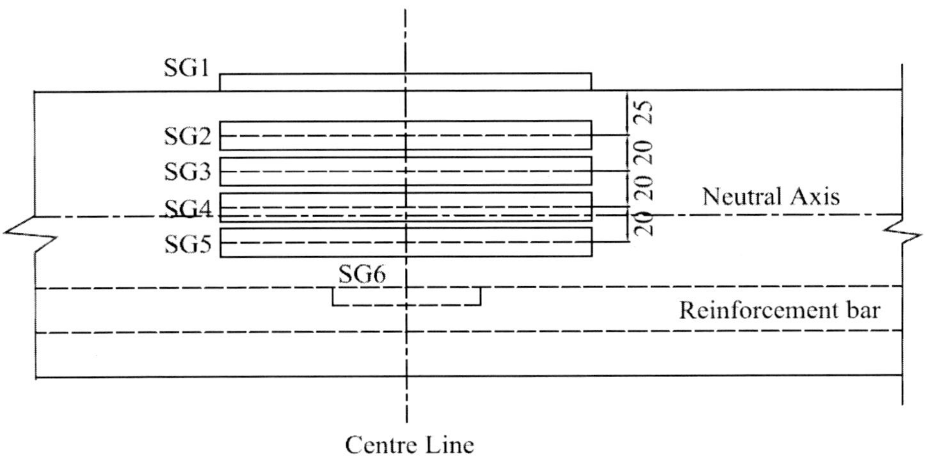

Figure 7.2. Locations of concrete and steel strain gauges [2].

Before placing the strain gauge, the reinforcement bars were cleaned using a wire brush to remove the rust occurring on the surface of the bars. The tensile steels were grinded to provide smooth and flat surfaces for the fixation of strain gauges (Figure 6.1, Chapter 6). The grinded surface was further smoothed with sand paper and later washed with acetone. The strain gauges were placed on the smooth surface using cyanoacrylate adhesive. The strain gauges used were TMA model: FLA-10-11 of Tokyo Sokki Kenkyujo Co. Ltd with resistant of 120 Ω and gauge factor of 2.0 (Figures 6.1 and 6.2 of Chapter 6). Finally, the strain gauges were connected to data logger with a lead wire by soldering on the terminal sheet. The strain gauges were protected by a layer of silicon to prevent damage during the casting of concrete. The strain values were automatically captured through a data logger connected to a computer through Visual Log software when the strain gauges were used.

For concrete strain measurement, an external grid of demec points was attached using superglue to the concrete surface as per Figure 6.5 (Chapter 6). The measurement of demec gauge is limited to a length of 200 mm (Figure 6.5 of Chapter 6) which is commonly used in the research [6]. Reading of demec gauge is recorded manually 2 times in an ascending and decending sequence to obtain a more precise reading. The precision of demec was $\pm 5 \times 10^{-6}$ under the controlled lab conditions. Theodolite was used to measure the rotation (Figure 6.6 of Chapter 6). A hand-held microscope with optical magnification of x40 and precision of 0.02 mm was used (Figure 6.7 of Chapter 6).

7.2.3. Dead and Imposed Loads

For panels with a thickness of 135 mm, the dead loads were the self-weight 2.7 kN/m^2 and 2.5 kN/m^2 for OPS hybrid LWC (concrete unit weight of 20 kN/m^3) and OPS LWC (concrete unit weight of 18.5 kN/m^3) respectively. The 25mm floor finishes of 0.6 kN/m^2, suspended ceiling and service facility of 0.6 kN/m^2 and partition wall of 1.0 kN/m^2 was allowed. The imposed load of 1.5 kN/m^2 was considered for private houses [7].

7.2.4. General Failure Modes of Panels

Table 7.7 shows the test results of the solid RC floor panels. The 28-day compressive strength of companion cubes using OPS LWC for panels S4, S5 and S6 achieved 27.60, 30.32, 26.50 N/mm^2 respectively. This OPS LWC however can be used in structural concrete elements which require minimum compressive strength 25 N/mm^2 in cast in-situ slabs [8, 9] or composite slab application [10]. The panels S1 and S2 achieved higher ultimate failure load compared to S4 and S5 respectively. Except for S1 and S4, all slab specimens exhibited ductile behavior expending large deformation prior to failure (Figure 7.3). The panels S1 and S4 with a steel ratio of 0.36% failed in brittle manner initiated by fracture of one or more of longitudinal tensile reinforcements accompanying a loud breaking sound where the panels collapsed into two halves (Figure 7.4). Typically, reinforced concrete slabs have relatively small tensile reinforcement ratios and are generally regarded as very ductile structural elements. This was however, not the case for panels S1 and S4 reinforced with low-ductility welded wire fabric. A similar observation was reported by [11, 12].

Table 7.7. Test results of solid floor panels [2]

Panel no.	Concrete type	28-day comp. st. of companion cube (N/mm^2)	Type of main steel	Panel failure mode
S1	OPS hybrid LWC	40.32	A10	Fracture of tensile reinforcement
S2		38.33	T10	Initial tensile yielding of steel
S3		38.90	T10	Initial tensile yielding of steel
S4	OPS LWC	27.60	A10	Fracture of tensile reinforcement
S5		30.32	T10	Initial tensile yielding of steel
S6		26.50	T10	Initial tensile yielding of steel

Figure 7.3. Typical flexure failure of solid floor panel reinforced with T10 [2].

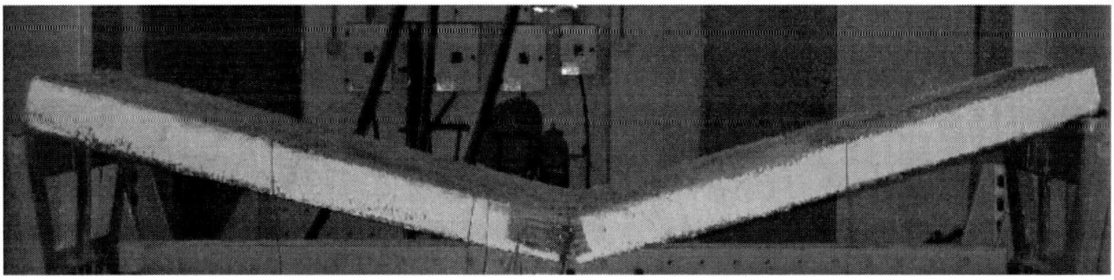

Figure 7.4. Failure mode of slab S1 and S4 reinforced with welded wire mesh, A10 [2].

7.2.5. Panel Weight

As mentioned, lightweight concrete is ideally suitable for precast construction application. Reduction in the self-weight of floor slabs therefore reduces the construction cost in handling, transportation and foundation as a whole. Table 7.8 shows the weight of panel specimens from S1 to S6. Comparing the weight of floor slabs made with normal weight concrete with density of 2,400 kg/m^3, there is an approximate weight reduction of 17 and 22.8% for OPS hybrid LWC and OPS LWC respectively. The weight of OPS hybrid LWC floor panels was higher than the OPS LWC floor panels by 7.4% due to the 30% partial replacement of OPS with crushed granite aggregate that possessed more than two times compacted bulk density compared to that of OPS aggregate. In the precast concrete industry, erection and installation of precast concrete floor panels made with OPS LWC become easier and labour costs can be minimized. Leading to further economic benefits in foundations, such reduction in dead weight is one important factor for earthquake-resistant designs. Therefore, these precast concrete flooring panels can be potentially used as alternative for the low-cost housing or plantation housing especially in the seismic regions.

Table 7.8. Solid floor panel and weight reduction compared to similar panel made with normal weight concrete, NWC [courtesy, author]

Slab No	Panel dimension, mm	Concrete Type and density	Panel weight, Kg	Panel weight, Kg	[1]Weight reduction %
S1	3,100mm(L)X1,000mm (W)X135mm	OPS hybrid LWC, 2,000kg/m^3	837	1,004	17
S2	3,100mm(L)X1,000mm (W)X135		837	1,004	
S3	3,100mm(L)X1,000mm (W)X175		1,085	1,302	
S4	3,100mm(L)X1,000mm (W)X135mm	OPS LWC, 1,850kg/m^3	774	1,004	22.8
S5	3,100mm(L)X1,000mm (W)X135		774	1,004	
S6	3,100mm(L)X1,000mm (W)X175mm		1,004	1,302	

Note: [1]If panel made with NWC of density 2,400kg/m^3.

7.2.6. Experimental Ultimate Load Capacity and Safety Factor

The load carrying capacity is of paramount importance to predict the safety, stability and durability of a structural element. It indicates the capacity of a member in withstanding its dead and imposed loads. Dead loads acting on slabs include the weight of the structure and any permanent material placed on the structure, such as tiles, ceiling and service facility, and walls. The imposed loads refer to those of magnitudes that may fluctuate with time due to their steadiness, movement and application caused by the weight of people, furniture, goods, and traffic etc. Table 7.9 shows the experimental ultimate and service load capacity of RC panels S1 to S6. The estimated service load was derived from dividing the experimental ultimate load by a factor of 1.7 [13]. The estimated Allowable Service Load (ASL) is equal to 3.7kN/m^2 area [Imposed load of 1.5kN/m^2) + Partition wall (1.0kN/m^2) + Services & Ceiling (0.6kN/m^2) + 25mm thick Screed/ or Floor tile (0.6kN/m^2)].

The panels S1 to S3 demonstrated a higher experimental ultimate load than panels S4 and S5. The addition of crushed granite in panels S1 to S3 increased the compressive strength of concrete and thus increased the load capacity. No significant change on experimental ultimate load capacity was noticed for panels S3 and S6 made with same thickness.

The type of steel reinforcement influenced the load capacity of precast solid floor panels made with OPS LWC. It can be seen that S1 reinforced with A393 and S2 with T10 achieved almost the same ultimate and service load capacity. However, S4 demonstrated 11% lower load capacity compared to its counterpart S5. It was suggested that the low-ductility welded wire reinforcement could not successfully transfer the stress to the surrounding OPS concrete matrix with lower tensile strength, which caused S5 to lack robustness and could not take up the energy to resist the load application. This was comparable to the findings [14].

Nevertheless, the safety factor of all solid RC floor panels ranged from 1.54 to 2.06 indicated that all panels were safe with high safety factors and thus OPS LWC is suitable for the residential dwellings such as low-cost housing.

Table 7.9. Experimental ultimate load and safety factor of RC panels [courtesy, author]

Panel mark (1)	Panel experimental ultimate load capacity (kN) (2)	Panel estimated service load = (2)÷1.7, (kN) (3)	Panel designed ASL (kN) (4)	Safety factor (3)÷(4)
S1	35.60	20.94	$3.7kN/m^2 \times 3.1m^2 = 11.47$	1.83
S2	36.03	21.19		1.85
S3	40.23	23.66		2.06
S4	29.94	17.61		1.54
S5	33.31	19.59		1.71
S6	42.15	24.79		2.16

Note: ASL = Allowable Service Load = 3.7kN/m2= [Imposed load of 1.5kN/m2) + Partition wall (1.0kN/m2) + Services & Ceiling (0.6kN/m2) + 25mm thick Screed/ or Floor tile (0.6kN/m2)].

7.2.7. Deflection Behaviour

Excessive deflection of a building floor may cause dislocation of the partition wall. Even in excessive deflection, the floor may not show cracks or damage anything, but large noticeable deflection can be psychologically disturbing to users. Figure 7.5 shows the load-deflection behaviour of six floor panels marked with S1 to S6. All panels showed the same deflection behaviour as steep gradient at initial stage for up to a load of about 12.5 kN except panel S4.

As the load increased, the change in slope occurred indicating the development of flexural cracking where the maximum tension stress at the bottom fibers was larger than the modulus of rupture of concrete. The slope of load-deflection curves remained fairly linear until the strain in steel yielded at this stage. Further load reduced the stiffness (EI) of the panels. However, to avoid damage, the linear variable displacement transducers (LVDT) were removed before reaching the expected ultimate load, hence the post yielding plateau of the load deflection curve was not captured. Nevertheless, it can be seen that the post yielding stiffness of all panels was influenced by the type of concrete, panel thickness and reinforcement provided. Both S3 and S6 panels showed almost the same behaviour even though they were made with different concretes. This was probably due to the thickness effect enhancing the moment of inertia (I) for contributing to the stiffness. The panel S4 made with welded wire mesh and OPS LWC showed the lowest performance due to a low modulus of elasticity of OPS LWC and low moment of inertia (I). As expected, the deflection of S3 and S6 was smaller than that of thinner slab specimens where their peak loads occurred at a lower displacement indicating a stiffer response.

Table 7.10 shows experimental loads with deflection limits and safety factors. The deflection limit is 3,000mm÷250=12mm. It is concluded that precast floor panels made with both OPS hybrid LWC and OPS LWC show better safety factors ranging from 1.22 to 2.66.

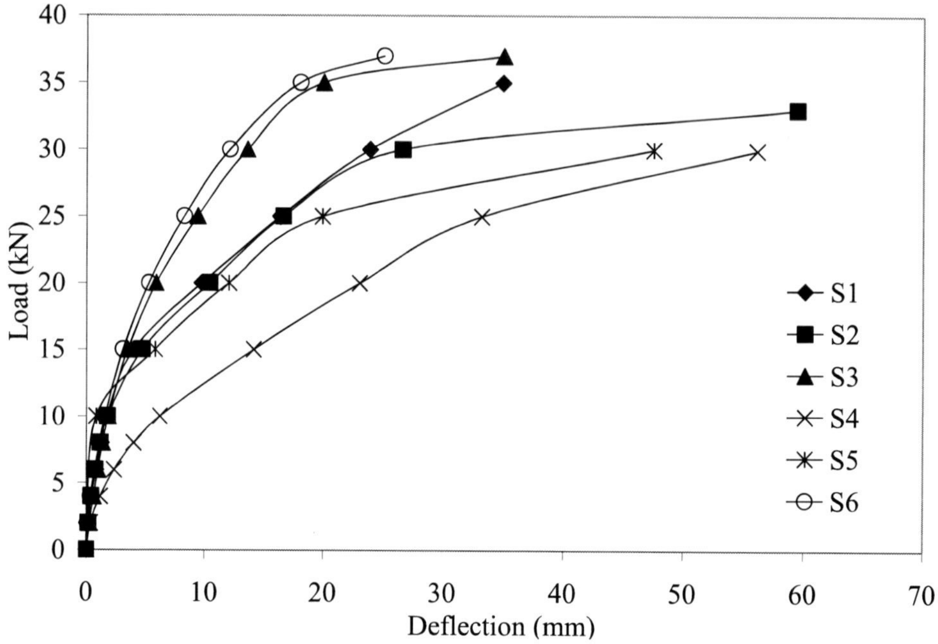

Figure 7.5. Deflection of different solid floor panels [2].

Table 7.10. Experimental load within deflection limit and safety factor of solid floor panels [courtesy, author]

Panel mark (1)	Panel experimental load within deflection limit (span÷250=12mm) (kN) (2)	Panel designed ASL (kN) (3)	Safety factor (2)÷(3)
S1	21.5	$3.7kN/m^2 \times 3.1m^2 = 11.47$	1.87
S2	21.5		1.87
S3	28.5		2.48
S4	14.0		1.22
S5	20.0		1.74
S6	30.5		2.66

Note: ASL = Allowable Service Load =3.7kN/m2 = [Imposed load of 1.5kN/m2) + Partition wall (1.0kN/m2) + Services & Ceiling (0.6kN/m2) + 25mm thick Screed/ or Floor tile (0.6kN/m2)].

Table 7.11 shows the experimental and predicted deflection at estimated service load. Based on BS 8110, the deflection of structural members must not exceed 12mm (span/ 250). The prediction deflection based on both BS and ACI codes was conservative.

Table 7.11. Experimental and predicted deflection at estimated service load [2]

Panel mark	Experimental deflection at estimated service load, Δ_{exp} mm	Predicted deflection (BS), Δ_{BS} (mm)	Predicted deflection (ACI), Δ_{ACI} mm	Δ_{exp}/Δ_{BS}	$\Delta_{exp}/\Delta_{ACI}$
(1)	(2)	(3)	(4)	(2)÷(3)	(2)÷(4)
S1	11.30	4.64	7.07	2.37	1.60
S2	11.90	4.69	7.58	2.54	1.57
S3	8.40	2.41	1.30	3.49	6.46
S4	18.10	8.76	15.48	2.07	1.17
S5	11.80	9.74	22.44	1.21	0.53
S6	8.20	5.66	7.90	1.45	1.04

7.2.8. Cracking Moment

Table 7.12 shows the comparison of theoretical and experimental cracking moments of panels S1 to S6. The experimental cracking moment was taken at the point where the sudden change of initial slope of the moment-deflection curve occurred. It is seen that the theoretical cracking moment overestimated the experimental findings by 155 to 458%. [15] reported that the use of f_r greatly overestimated the experimental cracking moments, and hence, a reduced value of about 55% of f_r should be used to predict the cracking moment of the beam structure with better accuracy. Nevertheless, for slab, a reduction of 20 to 35% of f_r was recommended. As anticipated, the M_{cr} of S1, S2 and S3 were found higher than their counterparts.

Table 7.12. Theoretical and experimental cracking moment of solid panels [2]

Panel mark	Theoretical cracking moment, Mcr (Theo) (kN m)	Experimental cracking moment, Mcr (Exp) (kNm)	Moment of inertia of cracked transformed section, Icr (×10-5 m4)	Distance of neutral axis from extreme compression fiber, x (mm)
S1	17.59	5.20	3.22	25.71
S2	17.59	5.20	2.28	21.04
S3	29.55	5.30	4.41	24.95
S4	12.76	5.00	6.14	35.35
S5	12.76	5.00	4.47	29.90
S6	21.44	5.10	8.81	35.71

7.2.9. Crack Width and Spacing

Cracks enhance the corrosion of steel reinforcement, surface deterioration, and its long-term detrimental effects. Hence, prediction and control of cracking, and crack widths are essential for reliable serviceability performance under long-term loading. The typical flexural cracking pattern on precast floor panels under pure bending is shown in Figure 7.6. The typical cracks formed were mostly vertical, perpendicular to the plane of the slab from tension face towards the neutral axis indicating failure in flexure mode. It was also worth noting that there was no horizontal crack observed, which further assumed that no relative slip failure occurred between the OPS hybrid LWC and OPS LWC, and the main reinforcing steel. In all specimens, the first crack occurred approximately at the midspan of the panel. When applied load was increased, new cracks either developed towards or away from the supports or extended from the existing cracks. As for panels S1 and S4 reinforced with welded wire mesh (A10 mesh), under increasing load, one of the flexural cracks within the moment region widened and further deepened into the compression zone which caused the brittle failure. A close examination of the cracked surface of the concrete cross-section revealed that the fracture path was not just affected by the bond strength between the OPS aggregate and the mortar phase but also the orientation of the OPS aggregate within the concrete matrix since the shape of OPS aggregate was elongated and flaky. As such, it was clearly observed that the crack path mostly travelled around, rather than through the OPS aggregate contributing to the failure in the floor panel.

Table 7.13 presents the crack details of solid floor panels S1 to S6. Generally the maximum no. of cracks occurred between two loading points. The factors influencing the compressive strength also influence the tensile strength. It is further evidenced from the average crack spacing between two loading points in the range from 92.19mm to 99.85mm for the tested six precast floor panels made with both OPS hybrid LWC and OPS LWC. No significant difference of total crack formations was noticed between two loading points made with two types of lightweight concretes. Also, the average distance of the first crack from the edge of the floor panel was not significant using two types of lightweight concretes. The distinctive difference between the mechanical properties of OPS aggregate and mortar phase imposed the tension concentration at the aggregate-mortar link, and it further accelerated the occurrence and spread of cracks. It was worthwhile to note that the average crack spacing was approximately less than one third of the secondary reinforcement spacing.

Table 7.14 presents the experimental and predicted crack width of panels S3 and S6 at the estimated service load. Crack widths have been observed to be primarily a function of the stress in steel, thickness of concrete cover, distribution of steel reinforcement, and relative distances from the neutral axis of reinforcing steel, and extreme concrete tension fiber [16]. Since the slab depth does not exceed 200 mm, the spacing between reinforcement 300 mm which was less than three times the effective depth, will ensure that the maximum permissible surface crack width in general will not exceed 0.30 mm at service load [4]. It can be clearly seen that the experimental crack width of panels S3 and S6 at the estimated service load was far below the predicted maximum crack width at both the reinforcement and surface level, and therefore they fulfilled the serviceability limit of cracking. Considering two of the permissible crack width values recommended [17] are 0.41 mm (dry air or protective membrane) and 0.30 mm (humidity, moist air, soil), precast floor panels made using either OPS hybrid LWC or OPS LWC can be used not only as flooring under dry exposure but also for damp or even wet areas with high humidity.

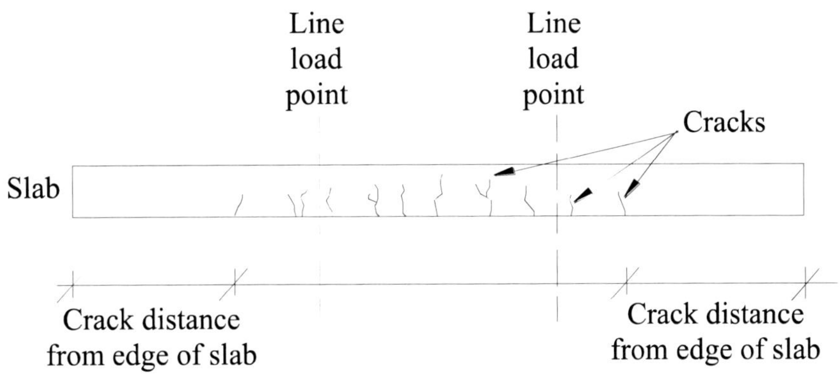

Figure 7.6. Typical flexural cracking patterns on slab under pure bending [2].

Table 7.13. Crack details of precast solid floor panels [2]

Panel mark	Max. no. of cracks between loading points	Average crack spacing (mm)	Average crack distance from end of floor panel (mm)
S1	9	94.29	737.50
S2	10	96.21	726.50
S3	9	99.85	697.50
S4	10	92.86	675.00
S5	10	93.99	785.00
S6	11	92.19	695.00

Table 7.14. Experimental and predictions of crack width for solid floor panels [2]

Panel mark	Experimental crack width	Predictions of crack width	
	Crack width at estimated service load (mm)	Max. crack width at the reinforcement level (mm)	Max. probable Surface crack width (mm)
S3	0.10	0.16	0.26
S6	0.04	0.16	0.27

7.2.10. Ductility

The ductility is the ability of a concrete section to crack and undergo a finite amount of rotation prior to failure. In order to mobilize the assumed load paths and redistribute the load resistance in a floor, the slab must possess certain ductility. This capability prevents the total structural collapse and provides protection to occupants of buildings. It is defined as the ratio of deflection at ultimate load, Δ_u to the deflection at yield load, Δ_y of the slab specimens, as calculated in Table 7.15. The floor panels, S1 and S4 reinforced with welded wire mesh which was less ductile compared to hot-rolled steel bars [18] failed in brittle mode, so ductility was not determined. According to [19], members with ductility ratio in the range of 3 to 5 has adequate ductility to withstand large displacements like earthquake forces. In view of that, all panels except S1 and S4 showed relatively good ductility. This was probably due to the shock absorbing characteristics of OPS aggregate as noticed with low values of AIV, ACV and AAV of OPS aggregate.

Table 7.15. Ductility ratio of solid floor panel made using OPS hybrid LWC and OPS LWC [2]

Panel mark	Yield stage		Ultimate stage		Ductility ratio,
	Moment (kNm)	Deflection, Δ_y (mm)	Moment (kNm)	Deflection, Δ_u (mm)	Δ_u / Δ_y
(1)	(2)	(3)	(4)	(5)	(5)÷(3)
S1	7.50	3.95	17.80	¬	¬
S2	15.00	26.63	18.02	145.38	5.46
S3	18.50	35.03	20.12	206.79	5.90
S4	12.50	33.23	13.80	¬	¬
S5	12.50	19.94	16.66	152.42	7.64
S6	18.50	25.11	21.08	203.18	8.09

7.2.11. Steel and Concrete Strains Characteristics

Table 7.16 shows the strain properties of main reinforcements used in precast RC floor panels made with OPS hybrid LWC and OPS LWC. Some of the strain gauges became inactive or damaged after the cracking of the panel specimens. It must be noted, therefore, the actual strain values were much higher than presented herein.

Based on Figure 7.5 the cracking load was defined as the point where the load-deflection curve changed abruptly. An estimate of the points of transition was made for specimens that did not show a sudden change but rather a gradual one. The cracking load occurred approximately at a load of 25 to 33% of the ultimate load. It confirmed through the sudden increase in the tension in steel with associated increase in the tensile strain in the reinforcement at the crack section in all panels, S1 to S6. As the cracked section of concrete was considered ineffective in resisting tensile stress, the tension component had to be contributed by the reinforcement.

As such, the reinforcement provided an internal restraint which further resulted in a significant increase in the strain of tensile reinforcement. The location of reinforcement relative to concrete extreme compressive surface had a degree of influence over its tensile strains. It was evidenced that panels S2 and S5 being 135 mm thick, the maximum tensile strain values were higher compared to panels S3 and S6, having 175 mm thickness. This is comparable to the study reported by [20].

Comparing panels which failed in brittle, the tensile strain values of panel S1 was lower than panel S4. As expected, panel S1 made with OPS hybrid LWC having higher 28-day compressive strength had higher tensile strength which could resist greater induced stress in the concrete tensile zone. The maximum tensile strain of panel S1 was 1,622 microstrain, approximately 20% lower, compared to panel S4 of 2,040 microstrain.

The concrete strain characteristics are shown in Table 7.17. It should be noted that the strain readings, in the concrete tension zone i.e., below the neutral axis where cracking was once initiated, were discarded as this would not represent the actual strain obtained. The maximum concrete compressive strain values of panels S3 and S6 were 5,672 and 5,880 microstrain respectively. These results show that OPS Hybrid LWC and OPS LWC are able to achieve their full strain capacity under flexural loadings. Several solid floor panels were left after such investigations as shown in Figure 7.7.

Table 7.16. Tensile steel strain of solid floor panels [2]

Panel mark	Cracking load (kN)	Tensile steel strain		
		Before crack (microstrain)	At yielding (microstrain)	Maximum recorded (microstrain)
S1	10.40	33	1,622	1,622**
S2	10.40	75	1,363	1,998
S3	10.60	116	872	872*
S4	10.00	95	2,040	2,040**
S5	10.00	50	1,430	1,430*
S6	10.20	108	897	984

Note:
* Maximum value recorded before strain gauge became inactive or damaged.
**Maximum value recorded before floor panel failed in brittle.

Table 7.17. Concrete strain at different depths of RC panels at service moment [2]

Slab No	Concrete strain at service moment (microstrain)					Max. recorded concrete compressive strain (microstrain)
	SG1	SG2	SG3	SG4	SG5	
S1	~	210	184	0	25	~
S2	~	184	59	0	2	~
S3	195	108	46	2	0	5,672
S4	~	487	25	0	0	~
S5	~	401	17	4	2	~
S6	343	174	59	0	0	5,880

Figure 7.7. Solid floor panels of 3m span stacked after investigation [2].

7.3. Structural Performance of Solid Floor Panels Made with OPS Aggregate during Short Lifting and Transportation

The precast RC structural components after casting are handled from demoulding to stacking, stacking to transportation, and transportation to erection for assembling the structure at the construction site. So, these sequences need lifting, transportation and erection. The short height lifting is required at the casting yard during demoulding, stacking and loading onto the truck. The term erection is usually used at the construction site where these structural members are lifted from the truck to final placement for assembling the full structure. The handling process generally includes lifting and transportation at the precast plant and, erection and placement at the construction site. These precast RC components sometimes are transported a few hundred kilometers from the precast plant to the construction site by road or waterway. At the precast plant, precast structural members are demoluded and stacked at a very early age usually less than a week. If not handled carefully, these operations may cause damage to the precast products. For application in precast RC floor panels, the minimum requirement of concrete compressive strength at the time of lifting or transportation is 10 – 15 N/mm^2 [21-23] and at service is 35 N/mm^2 (if reinforced) to 40 N/mm^2 [24, 25].

The structural investigation was conducted to verify the performance of newly made precast lightweight panels during lifting and short transport as required at precast casting yards during demoulding and stacking.

7.3.1. Materials and Experimental Investigation

OPS is hard in nature and does not deteriorate easily once bound in concrete and therefore it does not contaminate or leach to produce toxic substances [26]. Basically, it does not need any chemical treatment prior to concrete mixing. OPS concrete, a lightweight concrete (LWC) using OPS as coarse aggregate has been developed [27-31]. The 28-day compressive strength of this LWC is more than 28 N/mm^2 and the engineering properties are proven to be suitable for structural members [32]. However, both OPS LWC and OPS hybrid LWC (Tables 7. 1 to 7.4) was used to produce precast lightweight floor panels. The precast floor panels of 3,100 mm (L) x 1,000 mm (W) x 135mm (D) were cast with OPS hybrid LWC and OPS LWC, reinforced with 4T10 as main steel and 11T10 as distribution steel. The plywood moulds were used. Based on 28-day air-dry density of both concretes, the weight of both panels made with OPS hybrid LWC and OPS concrete were found to be 837 kg and 774 kg respectively, which was about 16.6 to 22.9% less compared to identical floor panels (1,004 kg) made with normal weight concrete.

Four lifting anchors known as Halfen-Deha spherical-head plate lifting anchors with additional 4T10 steel rods and 200mm long each (Figure 7.8) were installed at four locations in each precast solid panel. Two such anchors were placed at 0.25L from each end and widthwise, they were also fixed at 0.25W from each edge symmetrically to the load centre as recommended for flat slabs [23]. These lifting anchors were designed with a minimum safety factor of 3 for steel failure.

Figure 7.8. Typical arrangement of Halfen-Deha spherical-head plate lifting anchor with additional 4T10 steel rods of 200mm long each [2].

Figure 7.9. Precast floor panel being lifted by 5-ton crane for measuring strain parameters [2].

Three electrical strain gauges of model: FLA-10-11 were attached on the prepared surface of main reinforcement steel at mid-span and one-third point, and another on the surface of the additional steel bar above one of the Halfen-Deha anchors. Typical attachment of electrical strain gauges to steel rod is shown in Figures 6.1 and 6.2 (Chapter 6). Two electrical strain gauges of model: PL-60-11 were attached each on the longitudinal and lateral concrete surface at the depth of 115 mm from the panel's top surface. All the strain gauges were connected to a portable data logger from which the readings at a lift interval of 1 meter were recorded in a computer.

For lifting tests, the panels were lifted at the points of anchors by a 5-ton Top-Mech overhead crane (Figure 6.10 of Chapter 6) at the lifting speed of 5 m/min. The panels during lifting maintained a level condition (Figure 7.9). The lifting belt angle was fixed at 60 degrees with horizontal (Figure 7.9). The panels were lifted to a height of 5 m (Figures 6.10 of Chapter 6) and readings were recorded at every 1 meter lifting. Figure 7.10 shows how the precast floor panel is carried by forklift for measuring the strain parameters during transportation. The panel was supported on the forklift at 775mm (0.25L) from each end.

Figure 7.10. Precast floor panel of 3.1m (L)×1.0m(W)×135mm (T) with two support points being transported by 10-ton capacity forklift of brand Toyota, model No: 3FD100 [2].

Usually precast structural components during lifting for moulding imposes more stress than transportation and erection [33]. When the precast unit is first lifted out of the mould, the required force may be a multiple of the product's weight (Figure 7.11). This is caused by suction, adhesion and friction between the mould and the concrete unit [34]. Adhesion forces between the mould and the concrete vary according to the type of mould used [23]. For rough timber mould, the adhesion force is, q = 3.0 kN/m^2. The value of adhesion to the mould, H_a is thus calculated using equation (7.7):

$$H_a = q \times A \tag{7.7}$$

where,

A = area of contact between mould and unit being lifted

Tensile force at anchor during lifting from the mould, Z_a is calculated using equation (7.8) [34]:

$$Z_a = (G + H_a) \times \frac{\omega}{n} \tag{7.8}$$

where,

G = weight of precast unit (kN)

ω = spread angle factor and

n = number of load-bearing anchors (For an arrangement of four independent chains or continuous diagonal chain runs in slab lifting, only two anchors can be assumed to be load-bearing)

For transporting and launching, Z_a is calculated using equation (7.9) [34]:

$$Z_a = \frac{G \times \omega \times \psi}{n} \tag{7.9}$$

where,

ψ = impact factor

The forces acting on the lifting anchors are computed based on an impact factor of 1 as recommended for lifting speeds less than 90 m/min [34]. The spread angle of 60° for lifting is adopted here (Figure 7.9). The bond strength, f_b (N/mm^2) is calculated using equation (7.10):

$$f_b = \frac{F}{(\pi \times \varphi_e \times l)} \tag{7.10}$$

where,

F = applied force

φ_e = nominal anchor diameter (mm), and

l = anchorage embedded length (mm)

The applied force is determined using equation (7.11):

$$F = \frac{G \times f \times z}{n} \tag{7.11}$$

where,

G = total weight of precast panel

f = crane factor, 1.3 (for transport in precast yard) or 1.5 (for pitch/ transport/ install) recommended by [5]

z = angle of pull factor = 1/cos β. For β = 30° (angle of lifting belt with respect to vertical, Figure 7.9) z = 1.16

To simulate the transporting load on a panel from the casting bed to the storage place or onto the truck, the panel was transported on a 10-ton capacity forklift (Toyota Model No: 3FD100) at a speed of approximately 15 - 20 km/h on level ground (Figure 7.10). As for normal practice, the portal frame, tower crane, excavator or forklift are used depending on the size of precast plant. However, the strain was recorded at every 3 m of horizontal movement.

Figure 7.11. Demoulding of precast C-channel being done with extra manual force [2].

7.3.2. Surface Cracks

No surface cracks were observed, near any of the lifting points after lifting the floor panel. There was no bond failure between OPS LWC and the anchors. The value of adhesion to the mould was H_a = 9.3 kN. Therefore, tensile force, Z_a acting on each anchor in OPS hybrid LWC and OPS LWC panels during lifting from mould was 4.86 kN and 4.49 kN respectively, which was less than the permissible load (i.e., 10 kN) for components cast in 15 N/mm^2 concrete [34].

For transportation and launching, Z_a acting on each anchor in OPS hybrid LWC and OPS LWC panels was 4.86 kN and 4.49 kN respectively. In actual practice the maximum lifting speed can be as high as 120 m/min. Then Z_a acting on each anchor in OPS hybrid LWC and OPS LWC is 6.31 kN and 5.84 kN respectively. This is about 30% higher than the ones under lifting speed of 5.0 m/min, with a recommended impact factor of at least 1.3 (for lifting speed more than 90 m/min).

Bond strength, f_b of the OPS hybrid LWC and OPS LWC was 9.59 N/mm^2 and 8.88 N/mm^2 respectively. These values were approximately 24% to 32% of their respective compressive strength, and are comparable to the bond strength of OPS LWC [35], sintered pulverised fuel ash concrete [36] and aerocrete [37]. Similarly, no cracks developed on the panels due to loads induced during transportation.

7.3.2. Strain Distribution

During lifting, the universal-head lifting clutches rested against the concrete during the diagonal pull and therefore transferred the horizontal load through every anchor foot and its additional reinforcements directly into the concrete. Figures 7.12 and 7.13 show the strain distributions of OPS hybrid LWC and OPS LWC panels during lifting. No strain was recorded in the steel at the mid-span and the additional reinforcement of the lifting anchor for both panels made with OPS hybrid LWC and OPS LWC.

The strain in both steel reinforcement at the one-third point towards the unrestrained ends of both panels made with OPS LWC and OPS hybrid LWC showed the maximum value of 250.8 microstrain and 8.36 microstrain respectively. This was expected since the flexural strength and modulus of elasticity of OPS hybrid LWC is higher than OPS LWC [38].

Due to fact that OPS aggregate has a good shock absorbent nature as indicated by the aggregate impact value (AIV) and aggregate crushing value (ACV) compared to conventional crushed granite aggregate, the strains of OPS LWC both on the longitudinal and lateral concrete surface were generally smaller than that of OPS hybrid LWC.

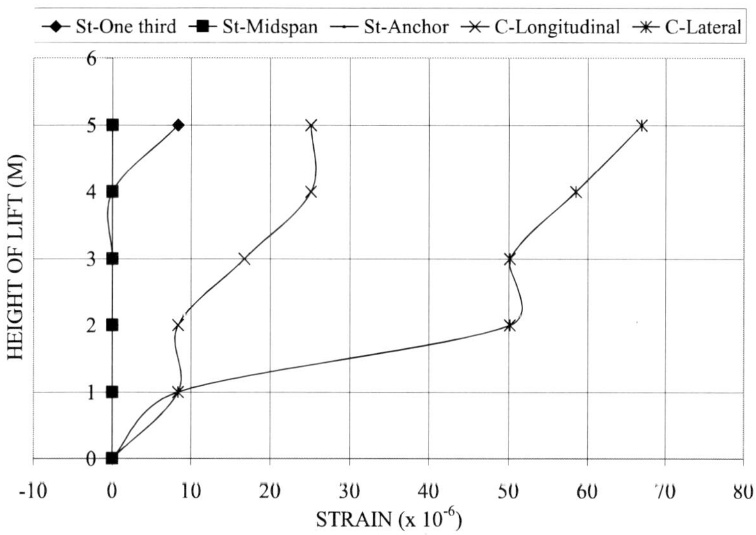

Figure 7.12. Strain distributions of OPS hybrid LWC panel during lifting [2].

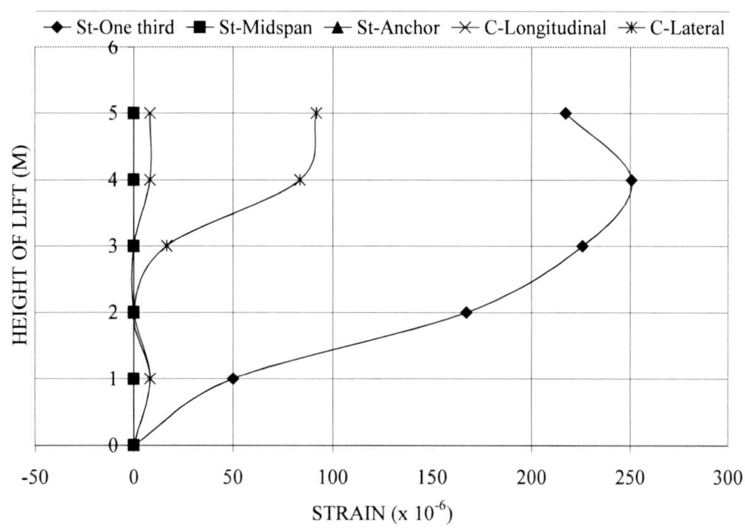

Figure 7.13. Strain distributions of OPS LWC panel during lifting [2].

Strain distributions of panels made with OPS hybrid LWC and OPS LWC during transportation with a forklift are shown in Figures 7.14 and 7.15 respectively. Similar to the effect of lifting, no significant strain is recorded in the steel at mid-span and anchor. Also, the steel at one-third point in OPS LWC showed relatively higher strain compared to OPS hybrid LWC. However, there was no identical pattern of strain distribution on the concrete surface. Forklift movement on the ground was not as smooth as lifting by crane. For OPS hybrid LWC, concrete strain at longitudinal and lateral surface showed a maximum strain of less than 200 microstrain, at horizontal distances of 12 m, which was less than 7% of the maximum allowable compressive strain in concrete at failure i.e., 3,000 microstrain [39]. In reality, the concrete strains will vary depending on vehicle speed, travel distance, terrain condition and transporting orientations. The effects of these parameters can be determined further through experimental investigation.

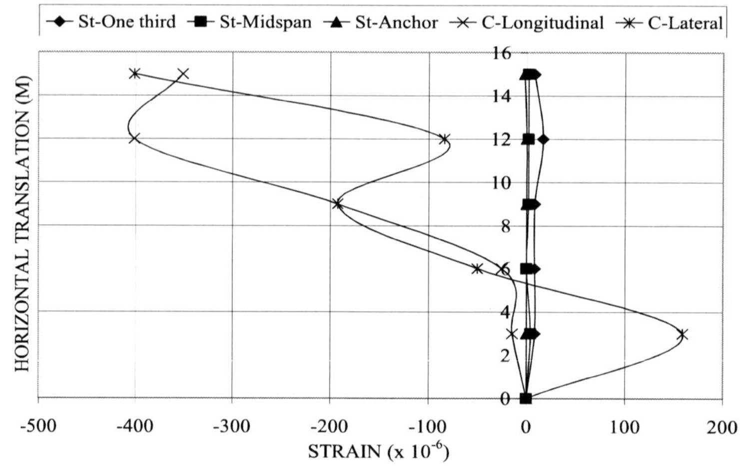

Figure 7.14. Strain distributions of OPS hybrid LWC panel during horizontal transportation by Forklift [2].

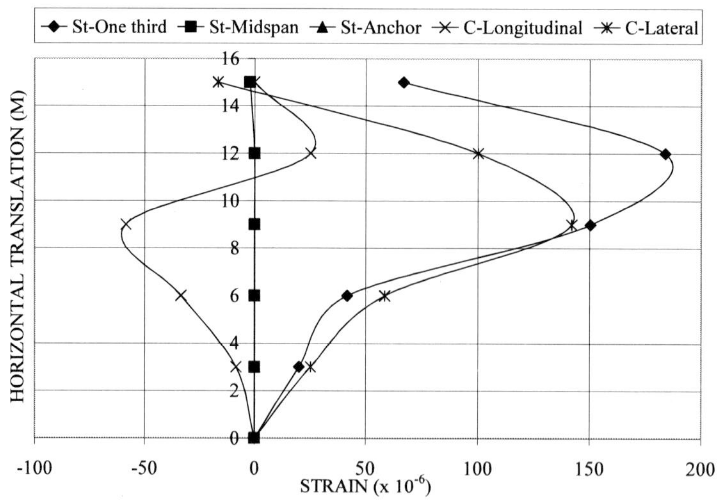

Figure 7.15. Strain distributions of OPS LWC panel during horizontal transportation by Forklift [2].

7.4. Flexural Behaviour of Semi-Precast Floor Panels with Different Toppings

Lightweight concrete using OPS as coarse aggregate has been investigated by many researchers with encouraging findings. The application of semi-precast floor panels for low-cost housing, the bonding and composite action between the base concrete slab and topping concrete using OPS LWC requires investigation. Very often, the topping concrete is essential for horizontal load transfer, flexural and shear strength, or simply to complete the construction. Such a study is needed in the case of remedying concrete deficiencies where extra concrete can be added by removing the non-structural screed before replacing it with a structural concrete topping.

Under the Industrialized Building System (IBS) concept, if the floor is made with full precast, it will have a large weight, which is difficult to transport and erect. As such, the semi-precast concept can minimize these difficulties. The semi-precast floor system is used in the construction industry worldwide. In semi-precast reinforced slab, base concrete is cast in a factory or offsite and topping concrete is placed after final placement of the slab panel. In this study of semi-precast technology, to reduce over dependency on natural stone aggregate, both base and topping concretes are made with conventional normal weight concrete and lightweight concrete. Two types of lightweight concrete known as OPS LWC and OPS hybrid LWC concrete were used (Tables 7.1 to 7.4).

This investigation discusses the semi-precast RC floor panels consisting of two layers of concrete where the concrete grade of the topping concrete was lower or equal compared to that of the base concrete made with OPS LWC or OPS hybrid LWC. Considering the application in low cost housing, especially rural and plantation housing, shear connector was not engaged as it would incur higher cost. Therefore, a simple, common yet economical bonding practice of applying cement slurry to the concrete substrate was adopted. When the base concrete reached the age of 7 to 18 days, topping concrete was cast upon it to form the slab with the tensile reinforcement of welded wiremesh, B196. Test parameters included cracks, deflection, moment and service load capacity, end rotation and strain distributions were considered. Results were also compared to semi-precast floor panels made with normal weight concrete.

7.4.1. Panel Fabrication and Test

A total of six floor panels of 3,100 mm (L) x 1,000 mm (W) x 50 mm (T) reinforced with BS4483 Ref. No. B196 (main and cross wire spacing 100 and 200 mm respectively) were cast with three mixes of concrete, namely OPS LWC, OPS hybrid LWC and normal weight concrete (Table 7.18, Figures 7.16 and 7.17). A concrete cover 20 mm was used for the main reinforcement. The panel surface was brushed with a coarse wire broom to remove laitance and scratched when concrete was in the plastic stage. They were cured using wet caissons for 7 days. Before placing the concrete topping, the base slab surface was cleaned thoroughly of contaminants and loose mortar or aggregate. The surface was also dampened, leaving it free of standing water. Immediately before placing the topping, cement slurry (mixture of cement with enough water) was spread onto the surface for better bonding purposes. At 7 to 18 days, different concrete toppings with 50 mm thickness were cast onto the base slabs to form full floor panels. No shear connector was used. When the topping concrete reached 28-days, the panels were tested under 2-line loading (Figure 7.18) at load increments of 2kN until failure. The distance between the loading lines was at 1,000 mm and loads were placed at the one-third positions of the 3,000mm span. The LVDTs capable of measuring maximum displacements of 100mm were used to measure the deflection at midspan and one third points of the panels. Three electrical strain gauges (Model: PL-60-11) were fixed (Figure 7.18) on the concrete vertical surface at depth 25 mm (strain gauge mark as SG1) and 80 mm (strain gauge mark as SG2). Two electrical strain gauges (Model: FLA-10-11) were attached on the prepared surface of a steel bar at midspan (strain gauge mark as SG3) and at one-third point (strain gauge mark as SG4). The LVDTs and strain gauges were connected to a portable data logger for recording the readings at preset load intervals until failure of the panel occurred. For all panels during testing, crack formations were marked on the vertical face on the panels. The end rotation of the panel was measured using a theodolite (Model NT-4D) with an accuracy of 1 second (Figure 7.18). The theodolite was positioned on the panel over the support point and a staff was placed some distance from the theodolite to observe vertical readings at every load increment.

Table 7.18. Floor panel concrete and strength properties [courtesy, author]

Panel mark	Base concrete and comp. st., N/mm^2	Topping concrete and comp. st., N/mm^2	Test at age (day)
S1	NWC, 58	NWC, 58	28
S2		OPS hybrid LWC, 38	28
S3		OPS LWC, 28	36
S4	OPS hybrid LWC, 38	OPS hybrid LWC, 38	28
S5		OPS LWC, 28	28
S6	OPS LWC, 28		28

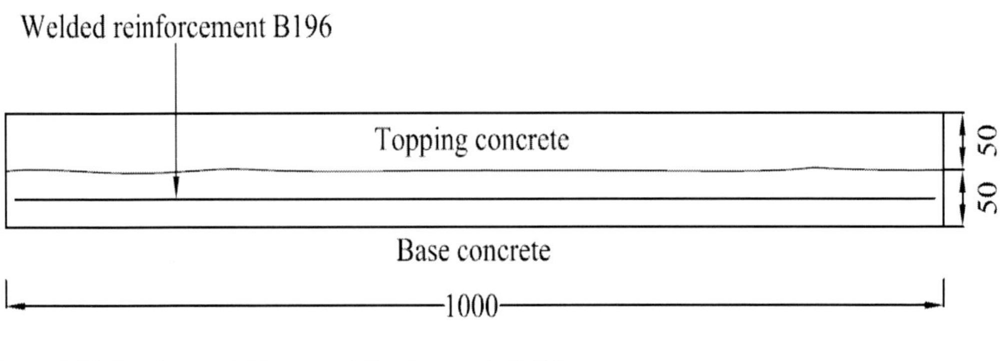

Figure 7.16. Semi-precast Floor panel fixed at m width [2].

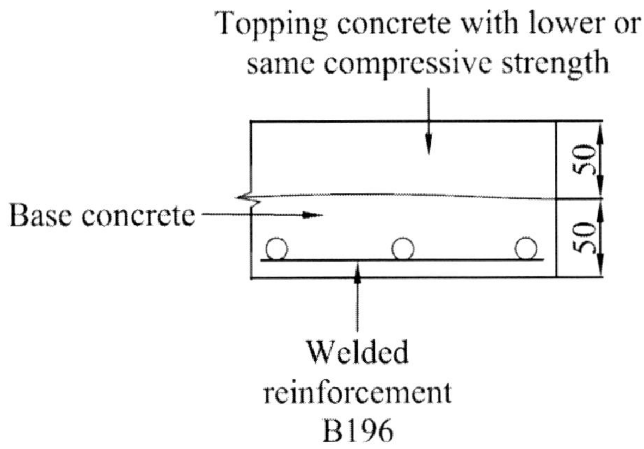

Figure 7.17. Cross section of Semi-precast floor panel [2].

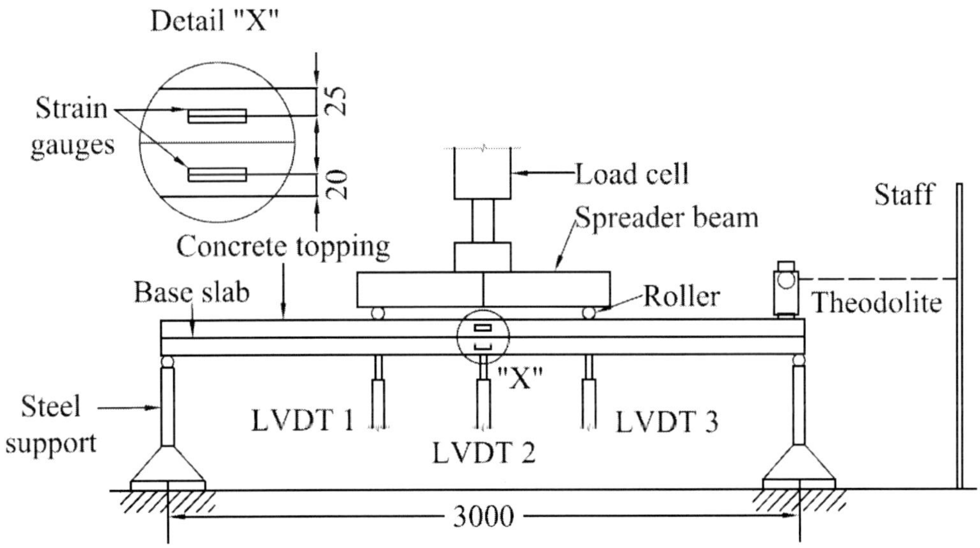

Figure 7.18. Experimental set-up for the slab specimen [2].

7.4.2. Failure Mode of Panel

Test results and observations are presented in Table 7.19 and Figures 7.19 to 7.24. Since all panels were lightly reinforced with a very small percentage of steel, the expected failure was to be brittle instead of in a ductile manner. When the tensile stress in the concrete exceeds the modulus of rupture, the concrete cracks and immediately releases the tensile force it carries. The stressed steel then absorbs this increment of load. When the area of steel provided is too small to carry this added force, the steel will break and a total rupture of the section will occur suddenly with a blasting sound. For panel S1, the failure was initiated by the partial rupture at the concrete interface within the pure bending region when the load reached approximately 6kN, then followed by the yielding of steel reinforcement under further load increment. It was due to the poor local bonding between the concrete topping and the base concrete. Differing from panel S1, panel S3 failed by steel-yielding which started at about load 4.8kN prior to partial rupture at the concrete interface at 7.5 kN. Such rupture did not occur in the other panels. It is worth noting that the cracking load increased with the use of normal weight concrete as the base concrete with topping concrete of higher compressive strength. In panel S1, the cracking load was approximately 57% of the ultimate load. The cracking load of panel S6 was only 21% of its ultimate load.

Table 7.19. Test observations and results [2]

Panel mark	Av. crack space (mm)	No of cracks between two load lines (max)	Crack load (kN)	Ultimate load (kN)	Failure mode
S1	204	3	6.0	10.51	Yielding of steel and partial rupture at concrete interface
S2	174	4	4.0	10.66	Yielding of steel and partial rupture at concrete interface
S3	232	3	4.8	12.10	Yielding of steel and partial rupture at concrete interface
S4	123	7	4.2	13.09	Yielding of steel
S5	154	5	4.0	13.55	Yielding of steel
S6	102	7	2.8	13.51	Yielding of steel

Figure 7.19. Failure pattern of Semi-precast panel S1 through yielding of steel and rupture at concrete interface [2].

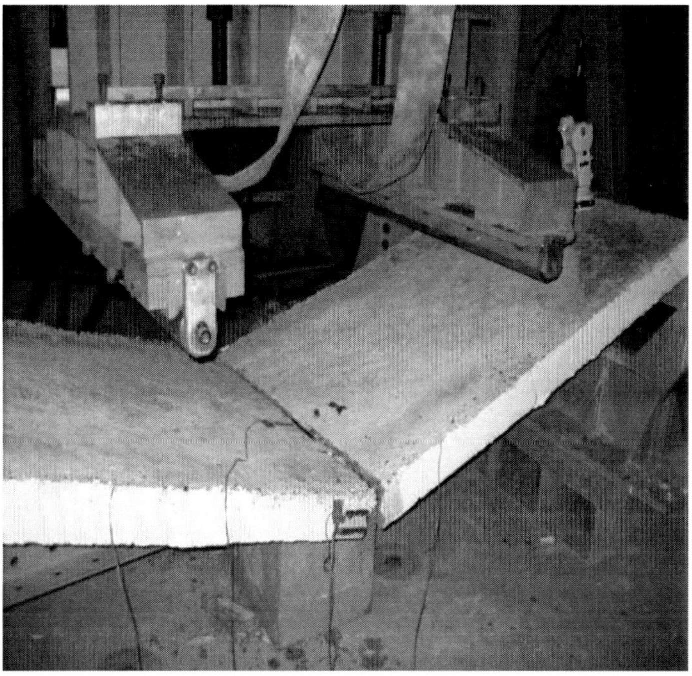

Figure 7.20. Failure pattern of Semi-precast panel S2 through yielding of steel and rupture of panel [2].

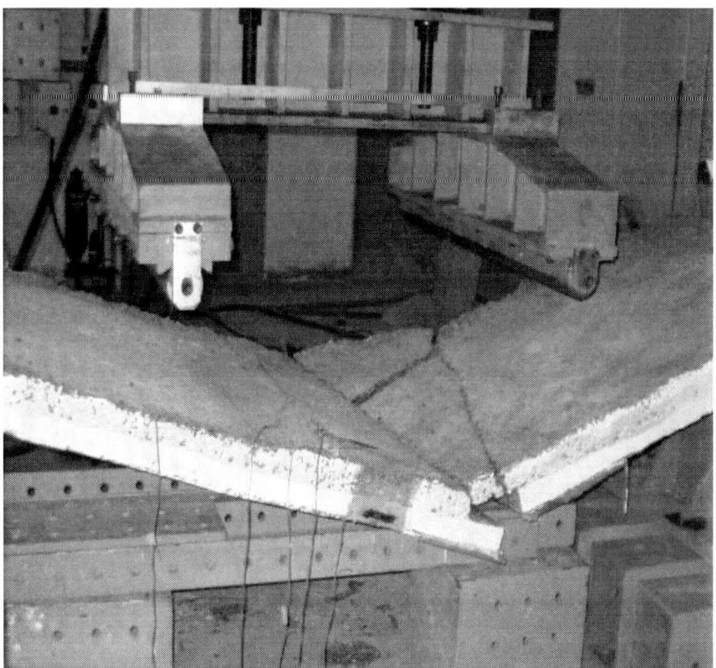

Figure 7.21. Failure pattern of Semi-precast panel S3 through yielding of steel and rupture of panel [2].

Figure 7.22. Failure pattern of Semi-precast panel S4 through yielding of steel [2].

Figure 7.23. Failure pattern of Semi-precast panel S5 through yielding of steel [2].

Figure 7.24. Failure pattern of Semi-precast panel S6 through yielding of steel [2].

7.4.3. Cracks

Almost vertical cracks were found for all floor panels. It was observed that the first crack occurred at the midspan of the panel. As the external load increased, additional cracks developed in the central region (bending region) of the span, then spread towards the supports. Also the initial cracks widened and extended toward the neutral axis and beyond. No horizontal crack was observed at the level of reinforcement, which indicated that there was no bond failure between the OPS or the conventional crushed granite aggregate and the steel reinforcement. Nevertheless, horizontal cracks were found within the bending region of panels S1, S2 and S3, at the concrete interface where the base concrete was made using normal weight concrete. For panels S4, S5 and S6, the concrete base made with either OPS LWC or OPS hybrid LWC showed rougher concrete surface texture which provided better bonding zone for topping concrete.

Table 7.19 shows the average crack spacing and maximum no. of cracks occurring at the base concrete between two line loadings. Since steel bars transferred forces through the bond continuously along their entire bonded length, cracks occurred at locations where the transferred shear stress from the longitudinal steel bar to the surrounding concrete exceeded the bond strength of the steel. Therefore, the crack spacing depended on the bond quality of the bar and slightly decreased as the load increased. Comparing panels S1, S2 and S3, the average crack spacing of panels S4, S5and S6 was smaller under higher ultimate load capacity. The crack in the OPS hybrid LWC and OPS LWC was mainly caused by the failure in the bond between the mortar phase and OPS aggregate phase, where the crack path went around the OPS aggregate.

The maximum no. of cracks between the two loading lines of panels S4, S5 and S6 was relatively generally higher than panels S1, S2 and S3. Generally, more crack formations are beneficial as they do not exceed the crack width specified in the standards.

7.4.4. Deflection

Figure 7.25 shows the load-deflection relationship of all panels. It shows that all panels were under-reinforced as their deflections were governed by both the concrete and steel properties. At a lower value of load, before the concrete has cracked, the floor panels deflected in a linear behavior which was primarily governed by the gross area of concrete. Though after the yielding of the reinforcement where there was a sharp increase of deflection with a small load increment, the panel collapsed abruptly at the breaking load and the final deflection could not be recorded. It was also seen that semi-precast floor panels with lightweight topping concrete showed less deflection compared to their counterparts. At load 12 kN, panel S5 with OPS LWC as topping concrete deflected 20.50 mm compared to that of panel S4 with OPS hybrid LWC as topping concrete i.e., 26.40 mm. One possible reason may be due to the fact that the panel did not fully behave monolithically without the use of shear connector, where the base concrete layer had to somewhat support the self-weight of the topping concrete. As expected, the panel with a concrete base made with higher concrete grade deflected less compared to that of the lower grade. The panel S6 deflected almost 8.5% higher compared to that of panel S4.

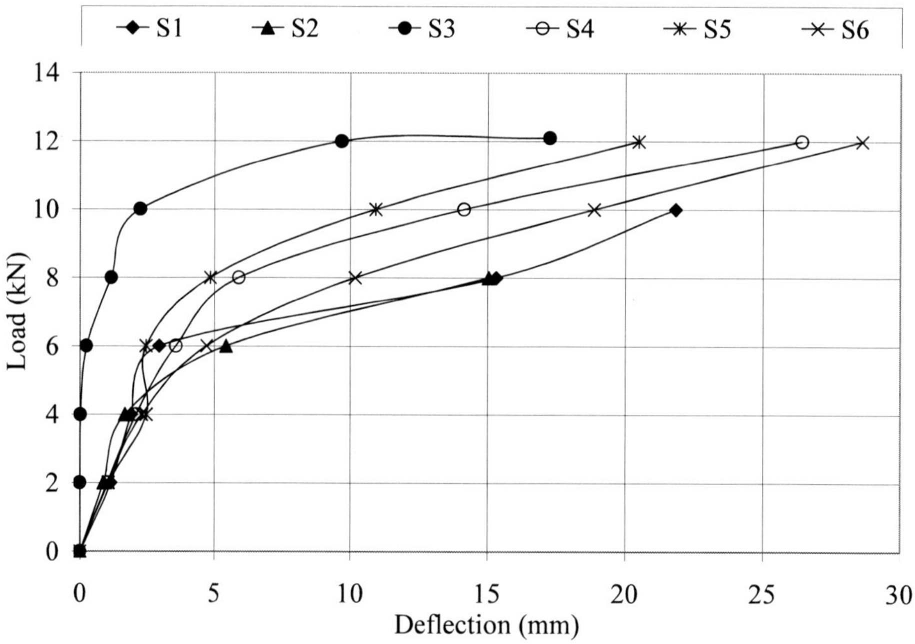

Figure 7.25. Load vs. deflection of panels S1 to S6 [2].

7.4.5. Moment and Service Load Capacity

Table 7.20 shows the service loads and moments of panels S1 to S6. These panels were investigated without considering the imposed load. The lowest and highest experimental moment and service load was obtained and shown in panels S1 and S5 respectively. The cracks occurred at the concrete interface in panels S1 and S3, reduced the composite action where the topping concrete as compressive zone was not able to resist the applied moment. On the other hand, panels S4, S5 and S6 with OPS LWC or OPS hybrid LWC base showed the higher moment compared with other panels. It indicated better bonding with the topping concrete even without the use of shear connector, compared to that of normal weight concrete base.

Table 7.20. Loads and bending moments [2]

Panel mark (1)	Design moment (kNm) (2)	Design service load (kN) (3)	Experimental moment (kNm) (4)	Estimated service load* (kN) (5)	Safety factor (4)÷(2)	Safety factor (5)÷(3)
S1	3.78	7.20	3.94	6.18	1.04	0.86
S2	3.47	6.60	4.00	6.27	1.15	0.95
S3	3.35	6.38	4.54	7.12	1.36	1.12
S4	3.15	6.00	4.91	7.70	1.56	1.28
S5	3.03	5.78	5.08	7.97	1.68	1.38
S6	2.91	5.55	5.07	7.95	1.74	1.43

*estimated service load = experimental ultimate load ÷ 1.7 [13].

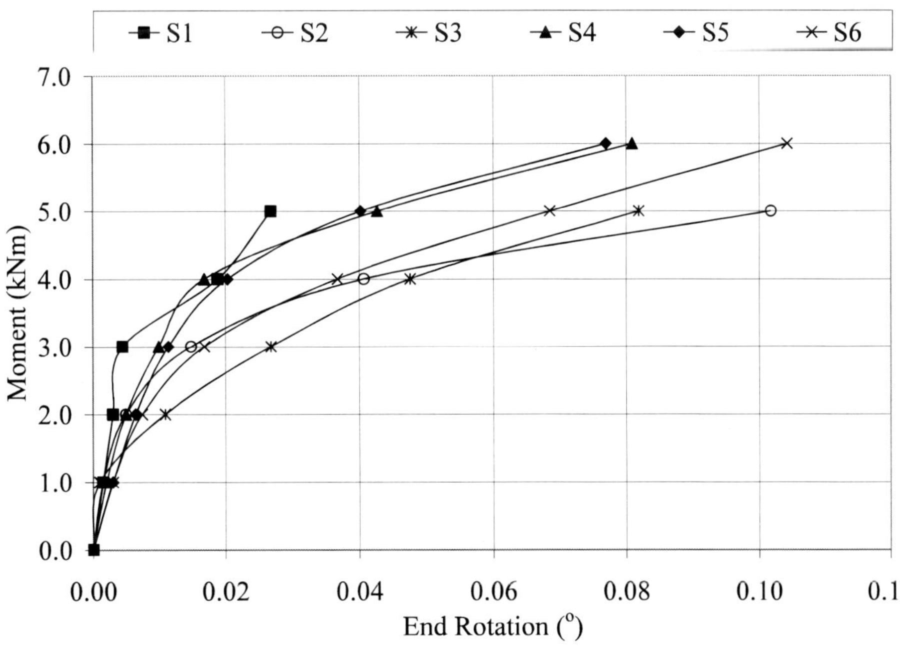

Figure 7.26. Moment – end rotation curves of panels S1 to S6 [2].

7.4.6. End Rotation

Figure 7.26 shows the curves of moment-end rotation for panels S1 to S6. This result is co-related well with the load-deflection relationship where the larger the end rotation, the more the deflection was. The finding of end rotation was of paramount importance as it points to the ductility of the structure, which is the ability of the structure to undergo bending deformation induced by earthquake forces, without rupture of the tension steel or crushing of the compression zone concrete. There was an apparent reduction in ductility with increased strength of floor panels where both the base and topping concrete were of the same strength. At 5 kN, the end rotation of panels S1, S4 and S6 was 0.027, 0.043 and 0.069 radians respectively. With lighter topping concrete, panel S5 was almost as ductile as panel S4.

7.4.7. Strain Distributions

The strain values of concrete and steel of panel S1 is shown in Figure 7.27. From the strain distributions of steel and concrete, it was found that there was a correlation between the strain values and the behavior of the panels. The section contained a relatively small amount of steel, where the steel reached its yield strength before the concrete reached its maximum strength. Taking the yield stress of B196 equals 485 N/mm^2, the yield strain was approximately 2400 microstrain. Typically, SG1 gave a higher concrete strain value than SG2 due to the flexural cracks that occurred at the extreme tension zone of the base concrete. Comparing to panels S1, S2 and S3, strain at the midspan of the steel (SG4) of panels S4, S5 and S6 was smaller, signifying a certain degree of composite action between the base concrete and topping concrete, where the topping concrete was able to resist the external moment effectively.

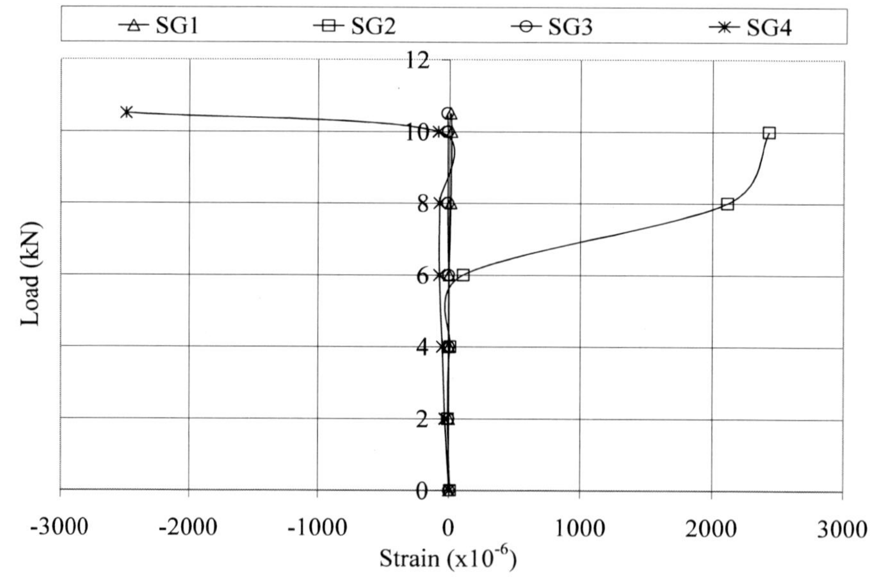

Figure 7.27. Typical concrete and steel strain distributions for panel S1 [2].

7.5. LIGHTWEIGHT PRECAST C-CHANNEL FOR BUILDING FLOORS

The exponential rate of population growth, development of industry and technology has created a need for more buildings particularly in urban areas. An example, Kota Kinabalu (KK), city of Sabah, Malaysia has land area of 351km^2 with the population density of 1,516/km^2 (≡303 houses/km^2, a family of 5 persons for one house). Concrete is the most consumed man-made material in the world. The floor slab needs more concrete compared with other structural components because it occupies a large space. It is estimated from the survey that a typical modern RC building needs concrete, about 44% for floor slabs, 31% for beams, 15% for foundation, 6% for columns and 4% for staircases.

The general concept of construction projects using precast concrete must be large enough to ensure sufficient profit to offset the initial capital cost. The precast concrete products are used in housing, institutional offices, shopping buildings etc. and must consider practical and economical aspects. The gravitational and seismic forces are proportional to structural weight. The lightweight building will enjoy having a reduced weight of structure.

However, due to the factors outlined above, it is an urgent issue to investigate the floor slab, particularly the precast reinforced floor slab. How can concrete be reduced while obtaining the highest structural performance both ultimate and service conditions over the conventional reinforced cast in-situ or precast floor panels? To meet the requirements as mentioned above, a full precast reinforced lightweight floor panel known as C-channel was developed. The C-channel is an award winning product a from the 20th International Invention, Innovation and Technology Exhibition (ITEX-09), Kuala Lumpur Convention Centre, 15-17 May 2009, Malaysia and Seoul International Inventions Fair (SIIF-2009), 3-7 December 2009, COEX, World Trade Centre, Seoul, South Korea.

The floor construction is the most time-consuming and costly activity particularly for a framed building, representing about 60-80% of the total in both cost and time [40, 41]. The floor slabs can be made of in-situ reinforced concrete or precast concrete. Generally, they are heavy in weight being made with normal weight concrete (NWC), and also, they require cast in-situ concrete topping which increases the cost for overall production. Prestressed elements which require a jacking system is very costly. Therefore, the proper selection of more efficient materials and improved design is required to reduce the cost of precast floor production. It is an objective on how a mobile prefabrication system can be developed to overcome fixed batching plants where conventional precast products are made. The mobile prefabrication system at temporary batching plants, next to project sites will reduce the transport cost of final products.

Also in line with the Industrialized Building System (IBS), precast floor systems particularly made with LWC have been known for the many advantages over the cast in-situ concrete slab. Lightweight concrete floor panels allow speedy erection thus shorten the overall construction time. The erection procedure, and in turn the design of the structure is very much affected by the weight and size of the individual components.

The structural behaviors of prototype precast reinforced concrete C-channels of 3, 4, 5, 6, 7, and 8 m spans have been investigated. The uniformly distributed loading (UDL) is the most common type of distributed load, in which a load of constant magnitude is applied along a length of a beam or an area of a surface. Not only so, uniform load is considered adequate for design purposes [42]. In the design of slabs, in addition to self-weight, a uniform imposed load is usually considered [43].

7.5.1. Industrialized Building System

Under the IBS, precast building elements are produced at a factory or off site, transported, erected and assembled to form a structure with little additional site work. It is not a new system worldwide but awareness is still not wide for many reasons. There are several advantages to using IBS in the construction industry as follows:

1) Cheaper total construction costs
2) Earlier access in trades
3) Faster rate of construction and early completion of project resulting reducing construction time
4) Less dependence on weather
5) Less industrial relations issues
6) Maintaining cleaner sites so reduction of risk at site
7) Minimization of site congestion as supply done based on demand
8) Minimizing storage area and vulnerability to damage for construction materials at site
9) On-time supply at site through co-ordination
10) Optimization of materials usage
11) Overcoming site difficulties through specialist skills and expertise
12) Precast elements being designed through test
13) Providing quality through dimensions, appearance, strength and durability
14) Reduction of construction materials at sites
15) Reduction of construction wastes
16) Reduction of site workers as products are made in factory or on off site
17) Reduction or elimination of formworks, props and scaffolding at site
18) Safer construction sites

7.5.2. Standard Dimension System and Tolerance

The standard dimension system and tolerances of precast components being very important parameters are adopted whether it is an open building system (OBS) or a closed building system (CBS). The standard dimension system is called modular coordination, ensuring compatibility between structural or non-structural building components as produced by different companies either in OBS or CBS. The dimension tolerances in precast units are limited to ±5mm in depth and width, and ±10mm in length [44]. The components with actual width provide construction tolerances and prevent overrunning due to cumulative errors. Hence, it is a very essential link among architects, design engineers, prefabricators and installers to work closely for the precast concrete project. The transportation and erection at site are also important aspects.

The moulds with adequate stiffness are required to maintain specified tolerances conforming to shape, lines and dimensions as specified in documentation for precast units. Assembling moulds before concreting is required to verify the dimensional tolerances as specified. The tolerances are required for precast products to specify for overall dimensions, thickness, bow, twist, flatness, squareness, size, location of openings etc. [45]. Moulds will be handled in repetitive operations through casting, demoulding and re-erection, so it should not deviate from the given tolerances. Moulds can be used twice daily depending upon concrete properties. Hence, it is recommended to verify the dimension-tolerance compliance of moulds at every 100 times of use [45].

7.5.3. Precast Concrete Floor Panels

In developing countries, precast flooring systems have not been accepted as another tradition in building construction. Less than 5% of the institutions of reputed higher education in Civil engineering have devoted courses on precast concrete [44]. This is one main reason why it has not gotten into the construction industry.

The precast concrete products, especially floor panels have several advantages over cast in-situ concrete. It provides both cost and design advantages over conventional cast in-situ concrete. It also speeds up the construction, minimizes on-site labour, is unaffected by weather, etc. Generally, the prestress concrete itself meets more than 90% of the demand for flooring [44]. The prestress concrete plant needs a high initial investment. Thus the high initial manufacturing cost discourages small and medium entrepreneurs from venturing into the prefabrication of precast concrete products.

Table 7.21. Conventional different types of precast floor panels [2]

Sl	Types of precast floor panel	Remarks
1	Flat slab	a) Use prestress and cast in-situ concrete for topping slab b) Need temporary propping during cast in-situ topping slab c) Need extra reinforcement for topping slab
2	Rib and Timber infill floor	
3	Hollowcore floor	
4	Spaced hollowcore floor	
5	Double Tee floor	
6	Concrete plank	a) Need cast in-situ topping slab b) Need extra reinforcement for topping slab
7	Concrete block system	
8	Channels slab	
9	Semi-precast (also called Half precast or Transfloor)	a) Use reinforced concreteNeed more reinforcement as compared with cast in-situ concrete b) Need temporary propping during casting of topping slab

At present, several types of precast floor panels are used in the construction industry. Flat slabs, rib and timber infill floors, hollowcore floors, spaced hollowcore floors, and double tee floors are used and they need topping slabs. Mostly, they need temporary propping during casting the in-situ topping slabs. Other types of precast floor slabs are concrete plank, prefabricated concrete block system and channels slab and they also need cast in-situ topping slabs. The semi-precast (also called half precast) floor panels can be made with reinforced or prestress concrete. The semi-precast floor slab needs more reinforcement as compared with cast in-situ concrete. Table 7.21 shows the conventional different types of precast floor panels.

7.5.4. Precast C-Channel Floor Panel

The conventional ribbed slab or channel slab is modified to form the C-channel as a single unit (Figure 7.28). It is not a pre-stressing panel. Therefore, a jacking system is not required. The soffit of C-channels for upper floors once covered with ceiling material is able to meet the acoustic properties that are useful for the reduction of noise transmission, and also to use for the service facilities along the span. The cost reduction for C-channels is achieved through the efficient material and optimum design. Therefore, the precast reinforced floor panel C-channel made of OPS solid waste or with reduced material of conventional aggregate can be a promising alternative over the conventional practicing concrete flooring systems.

Figure 7.28. A typical precast C-channel being lifted by Forklift [2].

Considering all the requirements as mentioned above, the C-channel as a precast reinforced concrete floor panel under IBS system has been developed, tested and verified with other full precast and semi-precast panels of similar sizes.

C-channels made with lightweight concrete (LWC), namely OPS hybrid LWC and OPS LWC have also shown a higher load carrying capacity both at ultimate and service conditions compared with solid and semi-precast panels (Sections 7.2 and 7.4). Figure 7.29 shows the 3m span C-channel cast with OPS hybrid LWC. Figures 7.30 to 7.33 show different preparations for 8m span C-channel with reinforcement placement in a steel-timber formwork, surface covering with plastic sheets after casting and lifting with two forklifts and finally placement of two units C-channels onto supports for structural investigation. In the construction industry, particularly in the housing sector, 3 to 5m floor spans are very common whereas 6 to 8m spans are used for other types of buildings like institutes, commercial building etc.

Figure 7.29. A 3m span C-channel cast in steel-timber mould using OPS hybrid LWC [2].

Figure 7.30. Reinforcement being placed for casting C-channel of 8m span using steel-timber mould [2].

Figure 7.31. Precast C-channel of 8m span being covered after casting with OPS hybrid LWC [2].

Figure 7.32. C-channel of 8m span made with OPS hybrid LWC being lifted by two forklifts [2].

Figure 7.33. Two units of C-channel each 8.0m span made with OPS hybrid LWC being placed on supports prior to structural investigation [2].

The two line loadings using an actuator of 1,000kN capacity was used for the structural investigation of 3m span C-channel (Figure 7.34), whereas Figures 7.35 to 7.39 show C-channels of 4, 6, 7 and 8 m spans loaded with UDL up to deflection limit.

The UDL was provided through layers of concrete block [420 mm (L) × 190 mm (H) × 125 mm (W)] or cement brick [200 mm (L) × 100 mm (W) × 65 mm (H)] with average weight of about 19.13 kg and 2.50 kg respectively in dry conditions. These blocks and bricks are easy to obtain, and they are used as load on test member with the space available. Their weight can be determined easily, and thus can represent the actual condition of the real load on the structure. The tests using UDL were conducted outside the laboratory, so rainwater or moisture changes in blocks or bricks were not considered.

Figure 7.34. C-channel of 3m span made with OPS hybrid LWC being investigated with two line loads using an actuator of 1,000kN capacity [2].

Figure 7.35. C-channel of 4m span made with OPS hybrid LWC being investigated with UDL of 21.95kN/m^2 (total load 87.8kN) up to the deflection limit. The author's lovely PhD student, C. H. Ng seen in investigation [2].

Figure 7.36. C-channel of 6m span made with OPS hybrid LWC loaded with UDL at 12.35kN/m^2 (total load of 74.10kN) up to deflection limit [45 with permission].

Figure 7.37. C-channel of 7m span made with OPS hybrid LWC being loaded with UDL at 13.95kN/m^2 (total load of 97.65 kN) up to deflection limit [2].

Figure 7.38. C-channel of 8m span made with OPS hybrid LWC being loaded with UDL at 19.80kN/m^2 (total load of 158.40 kN) beyond deflection limit [2].

Figure 7.39. C-channel of 8m span made with OPS hybrid LWC being loaded with UDL at 12.95kN/m^2 (total load of 105.20 kN) up to deflection limit. The author's lovely PhD student, C. H. Ng seen in investigation [2].

7.5.5. Performance Comparison of C-Channel with Solid and Semi-Precast Floor Panels

Table 7.22 shows performances of floor panels namely solid, semi-precast and C-channels of 3m span have been investigated with a design imposed load (DIL) of 2.0kN/m^2. The C-channel compared with solid and semi-precast floor panels has used less concrete thus C-channel has a concrete reduction of about 29.8%. At the deflection limit (span/250), C-channel has shown the largest suggested imposed load (SIL) of 6.46 kN/m^2 compared with two other panels. The C-channel has also shown the smallest average crack spacing of about 71.35mm as compared with the other two panels. Usually, large crack spacing allows formation of a wide crack width which is harmful for reinforced concrete structure as it may exceed the crack limit.

Different types of precast reinforced concrete floor panels using lightweight concrete (LWC) and normal weight concrete (NWC) were investigated as shown in Table 7.23. The 28-day compressive strength of normal weight concrete (NWC) and OPS hybrid LWC and OPS LWC were 58, 38 and 28N/mm^2 respectively. Similarly, the concrete unit weights were also 24, 20 and 18.5kN/m^3. OPS LWC showed the lowest unit weight. The C-channel made with OPS LWC showed the lowest panel weight of 5.80kN which can be transported and erected easily. In general, the C-channel showed about 30.95% concrete reduction compared with other panels. At the deflection limit, the C-channel showed the suggested imposed load (SIL) of 6.73 kN/m^2 which was higher than other panels. This indicated that the C-channel also showed the highest safety factors on an imposed load of about 4.49. The developed crack width at the estimated service load (ESL) in the C-channel was less than the limit crack of 0.3mm. The C-channel also showed the lowest crack spacing of about 59.80 mm which is beneficial to not allowing wider crack formation.

Table 7.22. Different types of precast reinforced floor panels investigated with design imposed load (DIL) of 2.0kN/m² [courtesy, author]

Type of cast	Solid panel	Semi Precast	C-channel[c]
Concrete type and grade	OPS hybrid LWC of 2,000kg/m³ density and G38		
Panel: 1.0m × 3.1m	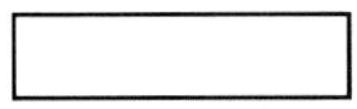	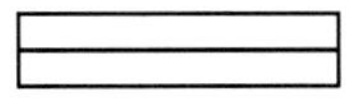	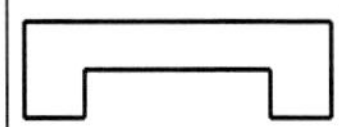
Panel thickness, mm	130	130	Flange=75, Rib=150
Main bar ratio, ρ (%)	0.24	0.36	Within limit
Distribution bar	T10-300	R6-250	Nominal
Shear connector	-	R6-120	a, b
Concrete volume, m³	0.40	0.40	0.28
Panel total weight, kN	8.0	8.0	5.60
Panel weight/unit area, kN/m²	2.58	2.58	1.81
Concrete reduction, %	0	0	29.8
28-day E, ×10⁴N/mm²	2.38		
Ultimate Load (UL), kN	66.13	64.75	65.20
Estimated service load (ESL), kN [ESL=UL÷1.7] [12]	38.90	38.09	38.35
DSL on span, kN	13.80	13.80	13.80
Deflection at ESL, mm	22.00	23.28	22.00
Determination of ASL at deflection limit (span/250=12mm), kN	21.00	19.75	27.19
ASL/unit area, kN/m²	7.00	6.58	9.06
Suggested Imposed Load (SIL) based on ASL, kN/m²	4.40	3.98	6.46
Safety factor =SIL÷DIL	2.2	1.99	3.23
Crack width at ESL, mm	0.06<0.3	0.06<0.3	0.08<0.3
Average crack spacing, mm	132.00	142.50	71.35

- Panel design based on 1.4×[dead load {self weight+1.0kN/m² (partition wall)+0.6kN/m² (service and ceiling)+1.0kN/m² (screed)} +1.6×[imposed load of 2.0kN/m²]
- Design service load (DSL) on span =13.80kN=3.0m×1.0m[1.0kN/m² (partition wall)+0.6kN/m² (service and ceiling)+1.0kN/m² (screed) + imposed load of 2.0kN/m²]
- [a]one-third of end span for both ends =double of middle-third span provided, [b]middle-third of span=nominal provided
- Allowable Service Load (ASL) = 3.0m (span)×1m (width) [Partition wall load of 1.0kN/m² +Ceiling and Service load of 0.6kN/m²+Screed load of 1.0kN/m² +Imposed load]
- Suggested Imposed Load (SIL), kN/m² =Allowable Service Load (ASL) - permanent dead load of (Partition wall load of 1.0kN/m² +Ceiling and Service load of 0.6kN/m²+Screed load of 1.0kN/m²)= ASL-2.6
- [c] lighter weight panel having advantage on transportation, erection and foundation

Table 7.23. Different types of precast reinforced concrete floor panels using lightweight concrete (LWC) and normal weight concrete (NWC) [courtesy, author]

Type of cast	Precast Solid panel as to conventional system[1]		C-channel[2]		
Concrete type, unit weight (kN/m^3) and 28-day Comp. st. (MPa)	NWC, 24 and 58	OPS hybrid LWC, 20 and 38		OPS LWC, 18.5 and 28	NWC, 24 and 58
Panel mark	1	2	3	4	5
Panel size: 1.0m wide × 3.1m long	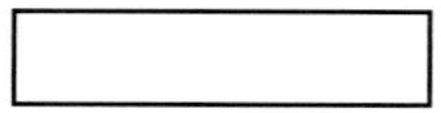	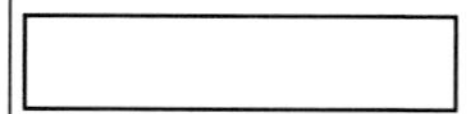	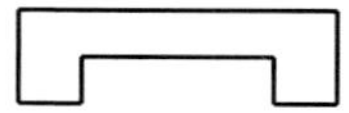	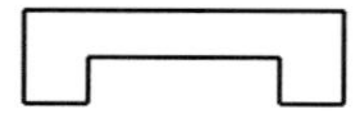	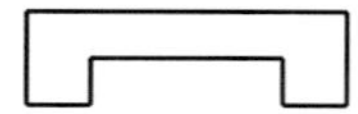
Panel thickness, mm	135	135	Flange =75 Rib=170	Flange =75 Rib=170	Flange =75 Rib=170
Coarse aggregate type	Granite coarse aggregate as identical coarse vol. to Panel 2	OPS hybrid coarse aggregate		OPS coarse aggregate	Same as panel mark 1
Concrete volume, m^3	0.42		0.29		
Concrete reduction (compared with panel mark 1), %	0	0	30.95		
Panel total weight, kN	10.08	8.40	5.8	5.37	6.96
Panel weight, kN/m^2	3.25	2.71	1.87	1.73	2.25
Main bar ratio, ρ (%)	0.23		Within limit		
Distribution bar	T10-300	T10-300	Nominal		
Shear connector	-	-	a, b		
28-day E, $\times 10^4 N/mm^2$	2.89	2.11	2.11	0.94	2.89
Ultimate Load (UL), kN	30.20	35.52	66.50	72.27	68.80
Estimated service load (ESL), kN [ESL=UL ÷ 1.7] [12]	17.76	20.89	39.12	42.51	40.47

Type of cast	Precast Solid panel as to conventional system[1]		C-channel[2]		
Design service load (DSL) over span, kN	4.50		12.30		
Maximum deflection at UL, mm	93	41.72	110	94.21	196
Deflection at ESL, mm	3.10	7.30	23.00	17.50	17.50
Allowable service load (ASL) at deflection limit (span ÷ 250=12mm), kN	24.25	24.40	28.00	25.00	30.00
ASL/unit area, kN/m^2	8.08	8.13	9.33	8.33	10.00
Suggested imposed load (SIL) based on ASL, kN/m^2	5.48	5.53	6.73	5.73	7.40
Safety factor =SIL ÷ DIL of 1.5 kN/m^2	3.65	3.69	4.49	3.83	4.93
Crack width at ESL, mm	0.16<0.3	Not recorded	0.18<0.3	0.09<0.3	0.21<0.3
Average crack spacing, mm	151.40	108.30	59.80	53.90	61.85

- [1]Panel design based on 1.4×dead load [self weight+0 kN/m^2 (partition wall)+0 kN/m^2 (service and ceiling)+0 kN/m^2 (screed)] + 1.6× imposed load of 1.5 kN/m^2
- [1]Design service load (DSL) over span =4.50kN=3.0m×1.0m[0 kN/m^2 (partition wall)+0 kN/m^2 (service and ceiling)+0 kN/m^2 (screed) and imposed load 1.5 kN/m^2]
- [2]Panel design based on 1.4× dead load [self weight+1.0kN/m^2 (partition wall)+0.6kN/m^2 (service and ceiling)+1.0kN/m^2 (screed)]+1.6× imposed load of 1.5 kN/m^2
- [2]Design service load (DSL) over span =12.30kN=3.0m×1.0m[1.0kN/m^2 (partition wall)+0.6kN/m^2 (service and ceiling)+1.0kN/m^2 (screed) and imposed load 1.5 kN/m^2]
- [a]one-third of end span for both ends =double of middle-third span provided, [b]middle-third of span=nominal provided
- Suggested Imposed Load (SIL), kN/m^2 =Allowable Service Load (ASL) - permanent dead load of (Partition wall load of 1.0kN/m^2 +Ceiling and Service load of 0.6kN/m^2+Screed load of 1.0kN/m^2)= ASL-2.6

7.5.6. Structural Performances of Precast C-Channels for 3 to 8m Spans

Figure 7.40 shows UDL capacities of all spans of C-channels within deflection limits. It shows that smaller spans have larger load capacity compared with larger spans. The C-channels of 3, 4, 5, 6, 7 and 8m spans (Table 7.24) were cast outside the laboratory and cured for 7 days with water-spraying. With regards to durability, such curing duration was also recommended for concrete made with OPS aggregate [47]. Later, they were exposed to the weather until the age of test with UDL up to the deflection limit (i.e., span/250). It must be noted that precast concrete panels are usually cast under protected shade at the precast plants. Three companion 100 mm cubes were also cast to determine the 28-day compressive strength of the corresponding C-channels.

The UDL, through a series of concrete blocks or cement bricks was applied on all C-channels. The blocks or bricks were arranged with gaps to avoid arch action. The LVDTs capable of measuring maximum displacement of 100 mm were positioned at midspan and at 300 mm from the end of span to determine the respective vertical deflections. The side of C-channels was white-brushed for crack detection. The demec gauges were fixed at mid span on top of the C-channel for recording the compressive strain of concrete. The electrical strain gauges were attached with rib main steel to record steel tensile strain.

The C-channels of all spans were designed with the imposed load of 3.0 kN/m^2 for all purposes including residential, institutional or educational buildings. At the deflection limit, the corresponding load on C-channels of 3, 4, 5, 6, 7 and 8 m spans was found to be approximately 18.00, 21.95, 16.60, 14.85, 13.95 and 12.95 kN/m^2 respectively. For C-channels, it was clear to notice the corresponding load capacity at the deflection limit reduced as the span increased. It is well understood that the structural element with a longer span must have a greater depth so as to provide adequate strength. Also, the self-weight of the element is directly proportional to its depth and so the ratio of the load carried to the self-weight per unit length of the element is less favorable for larger spans.

As C-channels longer than 3 m span tested up to the deflection limit and exhibited no evidence of failure, deflection recovery is further used as an evaluation criterion to determine whether the strength of the structure is satisfactory based on [48]. The 28-day compressive strength of the corresponding companion cubes was in the range of 22.15 to 23.58 N/mm^2.

To further assess C-channels of 3, 4, 5, 6, 7 and 8 m spans, their respective deflection recovery under a factored load was evaluated. It is noted that these C-channels were tested only up to the deflection limit (i.e., 12, 16, 20, 24, 28 and 32 mm respectively for 3, 4, 5, 6, 7 and 8 m span), and that their structural behavior at failure load remained unknown. It must be noted that the total load applied up to the deflection limit was higher than the total load required [49]. The deflection recovery of all spans of C-channels after one hour was between 87 to 98% which was more than 75% (Table 7.24).

Nevertheless, further investigation can be necessary to study the response of C-channels under failure load using uniform distributed loading. It must be noted that the ultimate behavior of C-channels tested under two-line load until failure is very much similar to panels made with normal weight concrete.

The existence of cracks induced by applied load affected the deflection recovery of C-channels as the degree of cracking affects the flexural stiffness. For C-channels of 8 m span, shear cracks observed was the main factor reducing the recovery percentage below 90%.

Table 7.24 shows the properties of precast C-channels made with OPS hybrid LWC having a shell thickness of 2.5 to 2.9mm for spans 4m and above. The panel weights from 3 to 8 m spans were in the range from 5.8 to 21.4kN which are suitable for transportation and erection. Also, C-channels made at 1m width can be accommodated with many units in trucks for transportation. The dead load of C-channels of 3 to 8m spans were from 1.81 to 2.61kN/m^2 area which makes building foundations less expensive as the total dead load from the C-channel floor will be less. The suggested imposed loads (SIL) at the deflection limit from 3 to 8m spans were in the range from 15.4 to 10.35kN/m^2. These were considered with high safety factors in C-channels over conventional design imposed load (DIL) of 3.0 kN/m^2 as used for design of different buildings. All spans of C-channels at ASL condition showed a crack width of less than the limit crack of 0.3mm. At ASL, the developed concrete compressive strains of all spans except 6 and 7m spans were from 520 to 1,230 microstrains which are less than the limit value of 3,000 microstrains. Similarly, rib tensile steel strains at ASL are in the range of 703 to 1,896 microstrains which are less than the limit value of 2,000 microstrains.

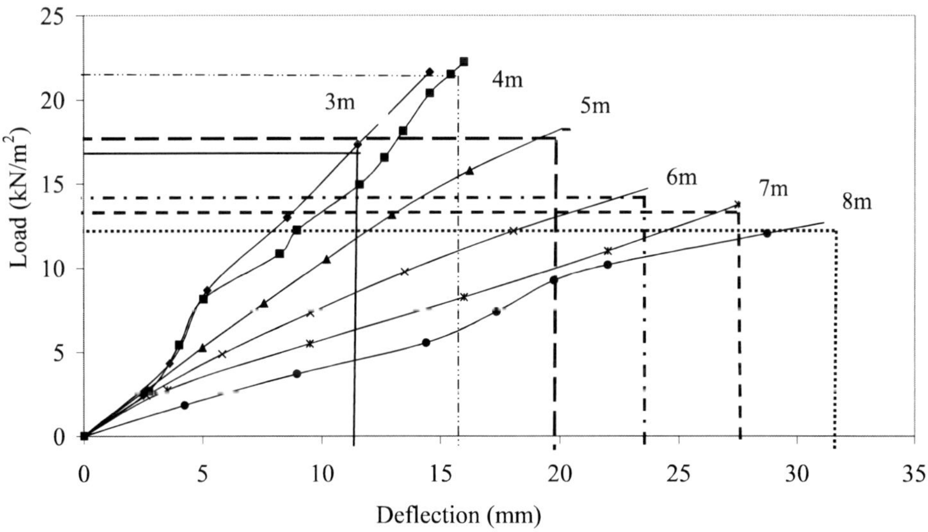

Figure 7.40. Load-deflection behaviour of C-channels of 3, 4, 5, 6, 7 and 8 m span made with OPS hybrid LWC loaded up to deflection limit (span/250) [46 with permission].

Table 7.25 shows general features of precast C-channels made with OPS as coarse aggregate partly or fully for LWC. The 28-day compressive strength of this new LWC using OPS as coarse aggregate was in the range from 20 to 38MPa. The overall panel depths were in the range from 170 to 350mm. All panels were made as reinforced concrete precast panels. This panel does not need any topping concrete once placed at the final position as side grooves will form a monolithic behaviour for C-channels. It also does not need any propping at final placement. At the site, it needs very little leveling for bearing on concrete beams which is determined as 100mm×100mm for each rib of C-channel. The service facilities like PVC pipes, electric wiring etc. are suggested to be installed at the soffit which is an open system, easy to repair or install. This C-channel is a promising alternative precast floor system as IBS in enhancing the speed of construction, providing immediate working space, having high imposed load capacity, cost effective materials, time saving etc.

Table 7.24. Properties of precast C-channels made with OPS hybrid LWC (OPS having shell thickness of 2.5 to 2.9mm for span 4m and above) [courtesy, author]

Sl	Description	Precast C-channel L=Channel length, (m) and Channels having 1m width, S=Span					
		L=3.2	L=4.2	L=5.2	L=6.2	L=7.2	L=8.2
		S=3	S=4	S=5	S=6	S=7	S=8
1	C-channel total weight, kN	5.8	8.6	11.4	14.6	17.8	21.4
2	C-channel weight, kN/m^2	1.81	2.05	2.19	2.35	2.47	2.61
	Compressive st. of companion cubes, MPa	23.40	23.07	23.58	23.04	22.66	22.15
	Total nos. of flexural cracks over span	16	18	27	31	28	45
3	Allowable service load (ASL) at deflection limit, kN/m^2	18.00	21.95	16.60	14.85	13.95	12.95
4	Suggested Imposed Load (SIL), kN/m^2	15.4	19.7	14.0	12.25	11.35	10.35
5	Design Imposed Load (DIL), kN/m^2	3.0					
6	Safety Factor=SIL÷DIL	5.13	6.57	4.67	4.08	3.78	3.45
	Required load for deflection recovery test as per Indian Standard, kN	15.95	27.79	35.39	43.10	50.95	59.35
7	Deflection recovery after one hour with removal of ASL, >75% as minimum	98	90	92	93	96	87
8	Concrete crack width at ASL, mm	<<0.3					
9	Concrete compressive strain at ASL, <3,500 microstrain	950	1230	520	-	-	960
10	Rib tensile steel strain at ASL, <2,000 miocrostrain	1,896	983	703	-	-	1,400

Suggested Imposed Load (SIL), kN/m^2 =Allowable Service Load (ASL) - permanent dead load of (Partition wall load of 1.0kN/m^2 +Ceiling and Service load of 0.6kN/m^2+Screed load of 1.0kN/m^2)= ASL-2.6

DIL = 3.0 kN/m^2 for several applications [7].

Table 7.25. General features of precast C-channel made with OPS hybrid LWC and OPS LWC [courtesy, author]

Sl	Description	Properties of C-channel
1	Panel width (m)	1.0 fixed
2	Span (m)	3, 4, 5, 6, 7 and 8
3	Panel overall thickness (mm)	170-350
4	Type of fabrication	Reinforced concrete
5	Concrete type and 28-day comp. st.	OPS hybrid LWC and OPS LWC, 20MPa to 38MPa
6	Type of steel used	Plain and deformed bars
7	Required Topping concrete	No
8	Required Propping system	No
9	Site Leveling work on Beam for Bearing	Very little, 100mm×100mm area/Rib
10	Service (PVC pipe, wiring etc.) facility installation	Open system, easy to repair/installation
11	Floor weight per unit area (kN/m^2)	1.81 to 2.61

Sl	Description	Properties of C-channel
12	Concrete saving compared with conventional cast in-situ NWC (%)	35.35 to 47.80
13	Speed of construction	Yes
Sl	Description	Properties of C-channel
14	Immediate working space	Yes
15	High load capacity	Yes
16	Factory produced	Yes
17	Fast erection	Yes
18	Cost effective (Materials and Time savings)	Yes
19	Structural deck with full diaphragm action for multistory buildings	Yes

All these are clearly demonstrating that C-channels, being the new member in the precast floor system saves the concrete, reduces structural weight, shows the highest structural performance in both ultimate and service conditions, is a cost effective and simple way of production. It does not need any topping concrete. Hence, the C-channels as developed here are highly suitable for floors in residential, educational institutions, offices etc. The moulds developed for C-channel production are considered the simple and mobile type, and it is suitable at temporary production plants next to the project site.

7.5.7. External Instruments on Precast C-Channels

External instruments were employed to investigate end rotation using LVDTs (Figure 7.41), vertical displacement at panel end with LVDTs (Figure 7.42), crack measurement with hand held microscope of x40 (Figure 7.43), and concrete strain using 200mm DEMEC gauge (Figures 7.44 and 7.45).

Figure 7.41. LVDTs used to measure the end rotation of C-channels made with OPS hybrid LWC under two line loadings using an actuator of 1,000kN capacity [2].

Figure 7.42. Set up of LVDT near supports for measurement of displacement while C-channels with loading [50].

Figure 7.43. Checking of crack formation on white painted surface of C-channels made with OPS hybrid LWC under two line loadings using an actuator of 1,000kN capacity. Author's lovely PhD student, C. H. Ng seen in investigation [2].

Figure 7.44. Concrete strain measurement at different rib-depth of C-channel using DEMEC gauge of 200mm gauge length [2].

Figure 7.45. Set up of DEMEC points at 200mm gauge length onto top surface of C-channels to measure concrete compressive strain during loading. Author's lovely PhD student, Doh Shu Ing seen in investigation [50].

7.5.8. Steam Curing

In conventional reinforced concrete practice, the striking of the side formwork from a concrete structure is normally carried out after 3 days of casting or if the compressive strength of the concrete reaches 10 to $15kN/mm^2$. However, for prefabrication, the striking of the formwork should be done as early as in 24 hours. In some especial cases, it prefers demoulding should be done within 6 hours after casting. Therefore, the construction industry will opt for a higher grade of concrete or implementation of steam curing. A higher grade of concrete basically means a higher cost of production. Thus steam curing is more favorable.

Due to high cost involvement in permanent steam curing systems, alternative methods of steam curing regarded as cheap and affordable can be implemented for the small scale prefabricator. This alternative method only requires (i) a boiler to generate steam, (ii) thermostat and thermometer to control the steam temperature, (iii) hose system to channel the steam to the products and (iv) covering sheet for optimising the curing effect of steam.

The steam curing for the precast industry may be of interest when 1-day compressive strength is the main consideration. This is beneficial, as the primary objective of steam curing is to obtain high early strength so that the precast concrete units can be demoulded early for repeated use of moulds.

The curing temperature will be a compromise between the rate of strength gain and compressive strength, as the compressive strength will be lower at the higher curing temperature [51]. At higher temperatures, the cement hydration rate increases quickly resulting in the formation of gel around each cement particle [52].

The microcrack and increased porosity during the heating stage may happen due to differences in the thermal expansion coefficient of concrete ingredients [53]. However, such deleterious effects may be prevented to some extent by the increased resistance to expansions when the steam curing of the concrete begins after a delay period equal to its setting time [54]. However, in this investigation using OPS hybrid LWC, the delay period was maintained as 3 hours since the final setting time of the OPC used was approximately 2 hours 40 minutes.

Table 7.26 shows the effect of steam curing on OPS hybrid LWC. The optimum temperature for steam curing for OPS hybrid LWC appears to be 65°C compared to 85°C even though LWC can be heated up to 82 and 88°C [55]. Under steam curing of 65°C for 16 hours, the 1-day compressive strength of OPS hybrid LWC increased 15% compared to water curing. However, it is suggested that the heat treatment application at lower temperature is more economical and energy saving [56].

Table 7.26. Compressive strength development of OPS hybrid LWC under different curing conditions [2]

Type of curing	Development of compressive strength			
	1-day	3-day	7-day	28-day
Water curing	17.30	29.50	35.97	38.53
Steam curing for 16-hours at 65°C	19.90	28.90	34.73	36.60
Steam curing for 16 hours at 85°C	19.20	27.20	34.17	35.72

7.5.9. Lifting Hook

Precast RC structural components are required during demoulding, stacking, transportation, erection and installation to a structural frame for the purpose of safe, effective and economic aspects.

They are able to demould from the casting bed once the concrete gains sufficient compressive strength. [33] has specified that the minimum concrete compressive strength for lifting and handling is 10 N/mm^2. During lifting from the casting bed, the precast RC components must remain stable as demoulding is usually done at an early age within 24 hours.

According to [45], the following parameters are required in order to determine the location and number of lifting hooks. They are:

1) Safe working load capacity of lifting hook
2) Weight of precast unit
3) Compressive strength of concrete at age of lifting
4) Shape of precast unit
5) Location of lifting hooks ensuring the safety of the unit although one lifting hook may cause failure
6) Position of any cut-outs and/or opening

The lifting hook, using steel rod is an additional element for the precast RC structural components. The following equations are used to determine the anchorage length of the lifting hook for precast C-channel.

The design anchorage bond stress, $f_b = \dfrac{F_S}{\pi \phi_C \ell}$ (7.12)

where

F_S = force in the bar or group of bars

ℓ = anchorage length, and

Φ_C = nominal diameter for a single bar or the diameter of a bar of equal total area for a group of bars

However, the anchorage length ℓ should be such that the bond stress does not exceed the design ultimate anchorage bond stress given by $f_{cu} = \beta \sqrt{f_{cu}}$ where β = a bond coefficient that depends on the steel bar type. β for deformed bars can be taken as 0.50 for tension or 0.63 for compression [57].

The design ultimate anchorage bond stress,

$$f_{cu} = \beta \sqrt{f_{cu}} \quad (7.13)$$
$$= 0.50 \sqrt{20}$$
$$= 2.236 \text{ N/mm}^2$$

Equate the anchorage bond resistance to the ultimate strength of the bar. Stress in the bar = 0.95 f_y.

$$f_b = 2.236 = \frac{stress \times \phi}{4\ell} = \frac{0.95 \times 460 \times \phi}{4 \times \ell} \quad (7.14)$$

$\ell = 48.86\, \Phi$

If T12 is used, then the anchorage length of the lifting hook ℓ is 586 mm.

Therefore, ℓ = 600 mm is taken into account (Figures 7.46 and 7.47).

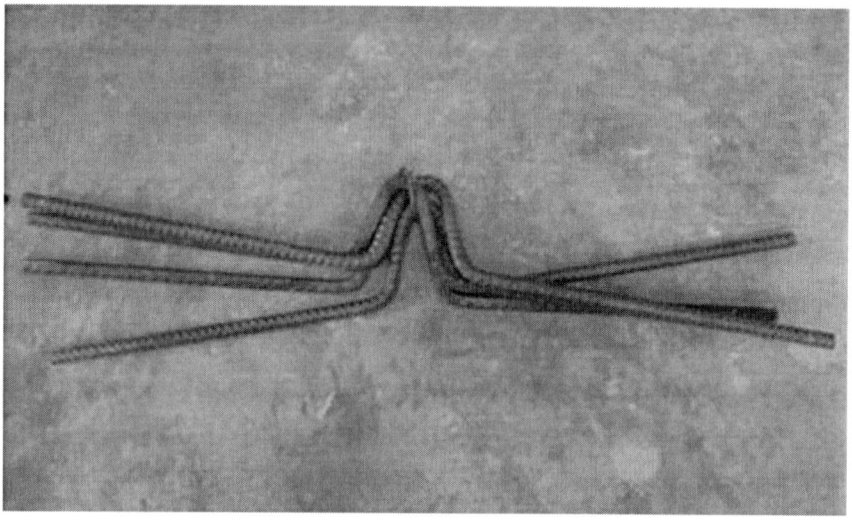

Figure 7.46. Lifting hook for 3m span of C-channel, anchorage length keeping with 600mm long for each end [2].

Figure 7.47. Typical pattern of lifting hook for larger than 3m span of C-channel while keeping anchorage length of 600mm for each end (depth of v-shape part of hook modified accordingly) [2].

7.5.10. Lifting Hook Placement in C-Channel

Four-points lifting for 3, 4 and 5 m spans, and eight-points lifting for 6, 7 and 8m spans were provided for precast C-channels (Figures 7.48 to 7.52). The lifting hooks for 3, 4 and 5m spans of C-channel were located at a distance of 0.208 L from the end of panel, where L is the full length of panel (Figure 7.48). Similarly, lifting hooks for 6, 7 and 8m spans were placed at 0.104L and 0.396L from each end of the panel (Figures 7.51 and 7.52). Panels were lifted using a forklift through steel chains. The forks of the forklift acted like the lifting spreader to lift the panels during demoulding, short transportation, stacking etc. However, the exposed part of the hooks were cut-off after installation of C-channels.

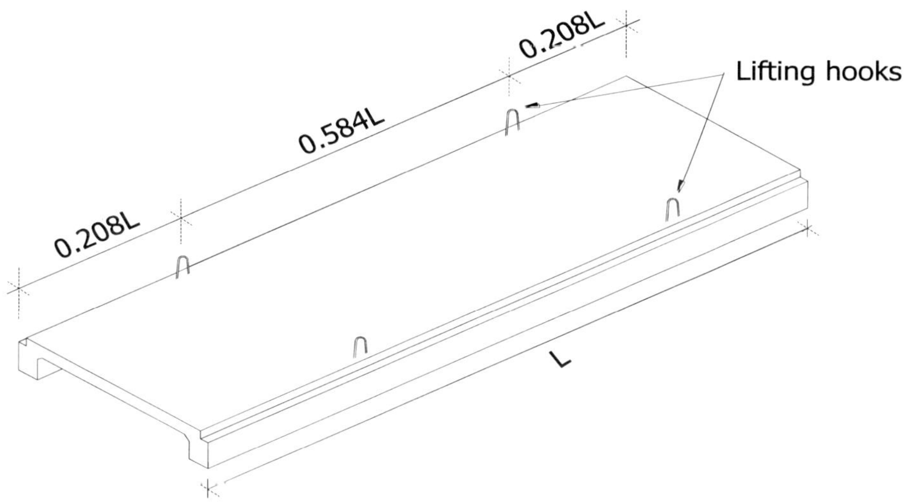

Figure 7.48. Position of lifting hooks for C-channels of 3, 4 and 5m span [2].

Figure 7.49. A 5 m span C-channel being stacked using Forklift with steel chains connected at four-points to lifting hooks [50].

Figure 7.50. A 5 m span C-channel made with OPS hybrid LWC being placed for platform formation by a truck mounted crane connected with steel chains at four-points to lifting hooks [2].

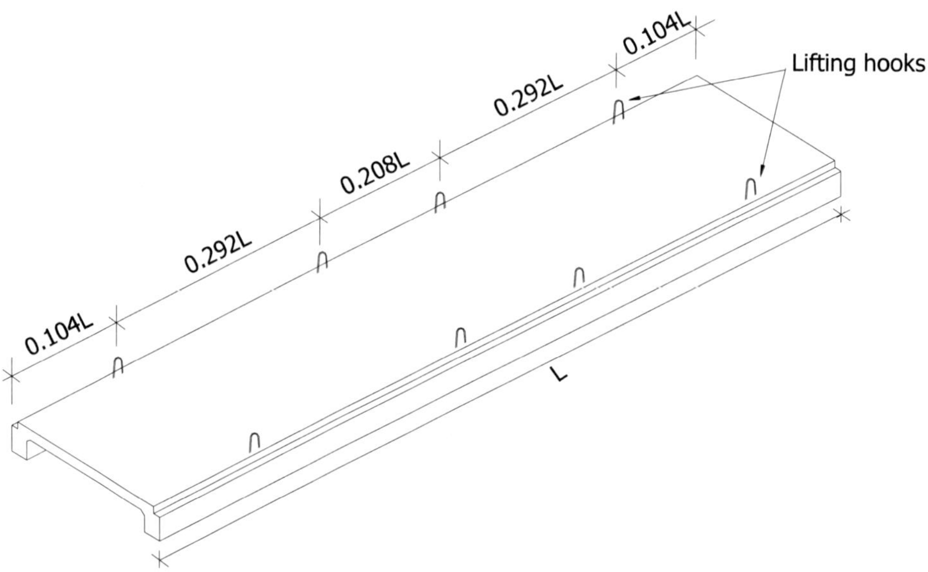

Figure 7.51. Position of lifting hooks for C-channels of 6, 7 and 8m spans [2].

7.5.11. Stacking of C-Channels

The proper stacking of precast floor panels are normally done at the production plant, shipment and construction site to minimize the storage space and to avoid excess stress development which is not considered in design. Bearers were placed vertically within 300 mm from the lifting hooks (Figures 7.53 and 7.54). The stacking usually is done on firm hardcore and level ground to avoid longterm creep development due to incorrect stacking. The height of stacking is another consideration so a skilled person can reach to pass the lifting chains or slings around the components [58]. The polythene can be used to wrap precast elements as to protect against stain and damages.

The careful inspection on all C-channels while in stacking was done to notice any crack formation. However, it did not show any defects so this kind of stacking for C-channel was suitable.

Figure 7.52. A 8 m span C-channel made with OPS hybrid LWC being connected with steel chains at eight-points to lifting hooks [2].

Figure 7.53. Stacking of precast C-channels using Forklift while maximum stacking height not exceeding twice the width of panel [50].

Figure 7.54. Stacking of precast C-channels with bearers while maximum stacking height not exceeding twice the width of panel [2].

7.5.12. Diaphragm Action

The overall structural performance of precast floor system depends on the integrity of the joint among all structural members such as floor panels, beams and columns. Floor spans of longer than 6 to 10m are designed for carrying gravity load and diaphragm action due to lateral forces like wind or earthquake [59]. Diaphragm action is possible to achieve with the addition of 50 to 75 mm topping concrete or integral shear and tie reinforcements [60]. The longitudinal edge profile of adjacent precast floor panels allows the placement of cast in-situ concrete for ensuring horizontal diaphragm action [59]. He has also elaborated that certain types of precast floor units are not capable of carrying horizontal forces in the floor diaphragm for three main reasons as follows:

1) Too thin flange of precast floor having 50 to 75 mm thick
2) Ends of precast floor units not tied to supporting structure to transmit diaphragm forces to the vertical bracing elements, and
3) No horizontal shear transfer mechanism between individual floor units.

The rib depth of a C-channel allows horizontal shear transfer between individual C-channel units as in Figure 7.55. Also, the gap provided between two C-channels is enough for proper placement of cast in-situ concrete.

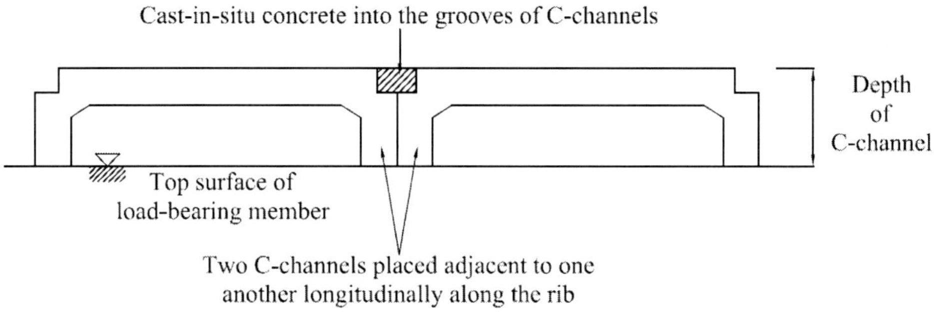

Figure 7.55. Rib depth permitting horizontal shear transfer [2].

7.5.13. Joint Design for Precast C-Channels with Pull-Down Mechanism

The main function of a joint is to prevent relative displacement between precast floor units. The joint size, material type used, and the design of joints are of important parameters to ensure monolithic behaviour between precast floor panels. [61] has pointed out that the final appearance of the joint must satisfy visual, fire and environmental requirements.

The C-channels, like other precast floor panels are placed side by side along their length, the dispersion of loads between panels has to be provided by the shear resistance through the rib depth. The flange groove along the length is filled with cast in-situ concrete of structural grade hereby known as joint, as shown in Figure 7.55. The advantage of this joint is simplicity in operation at the site. The effectiveness of this continuous simple joint as a flexural joint depends on the concrete's strength and reinforcement. At the connection, the protruding stirrup acts together with the cast in-situ concrete as reinforced concrete section where concrete in compression and steel in tension.

Precast concrete joints containing plain concrete are weaker than reinforced concrete. In high seismic zones, a reinforced precast joint is essential [62]. It is to mention that the concrete cover of the stirrup is approximately 25 mm, a very common practice in RC design. In the case, where cast in-situ concrete is placed against another precast or cast in-situ concrete element, the minimum concrete cover may be reduced to a value corresponding to the requirement for the bond provided [63]. The steel rod of 10mm dia. is recommended in this C-channel joint as longitudinal and cross reinforcements so cast in-situ concrete can be compacted with required reinforcement cover (Figure 7.56).

The purpose of the joint for precast floor units is to transmit forces between structural components and to provide stability by preventing relative displacements between units. The joint integrity for C-channels was investigated through the structural connection of four C-channels for 5m spans so as to form a monolithic floor as shown in Figures 7.57 to 7.62. No temporary propping was used during placement as C-channels are structurally efficient in terms of flexural rigidity and stiffness (Figure 7.50).

The monolithic behavior between precast C-channels was required with the intention to avoid topping concrete. Grooves of C-channels were connected through the pull-down mechanism using proper reinforcements that were fastened to the protruding rib stirrups between two adjacent C-channels. No welding was done for steel bars. This mechanism was to avoid unequal deflections of C-channels due to application of line load or point load. Prior to casting of in-situ concrete, the grooves were cleaned of any debris, loose mortar, dust, and dampened. The design calculation of the pull-down mechanism is shown below.

The total imposed load acting on 0.84m² area with load rate of 3.0kN/m² (Table 7.24) is 2.52kN. Thus there are four pairs of R10 mm bars along span resist 2.52 kN for 0.84 m² loaded area.

Thus, the load carried by one pair of R10 mm bar = 2.52÷4 = 0.63 kN.

However, one pair of R10 mm bars consists of two R10 mm rods. Thus each rod supports a load of 0.315kN (0.63÷2 = 0.315 kN).

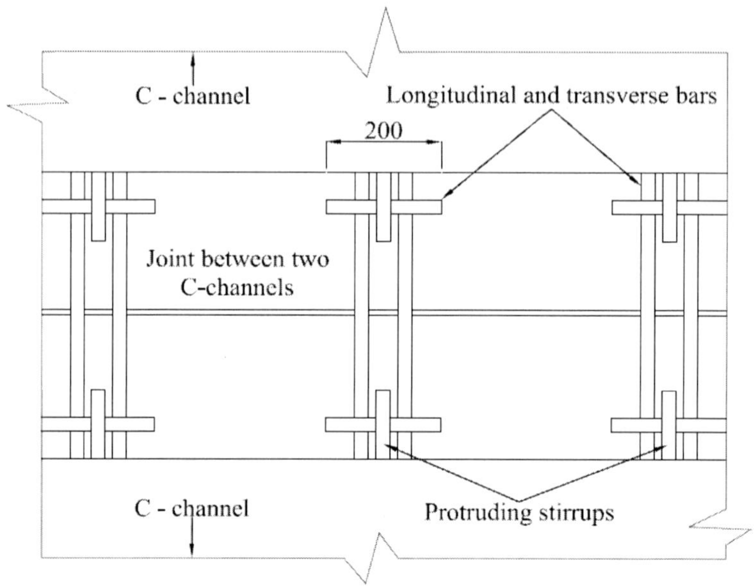

Figure 7.56. Schematic joint using steel rod as longitudinal and cross reinforcements with protruding stirrup [2].

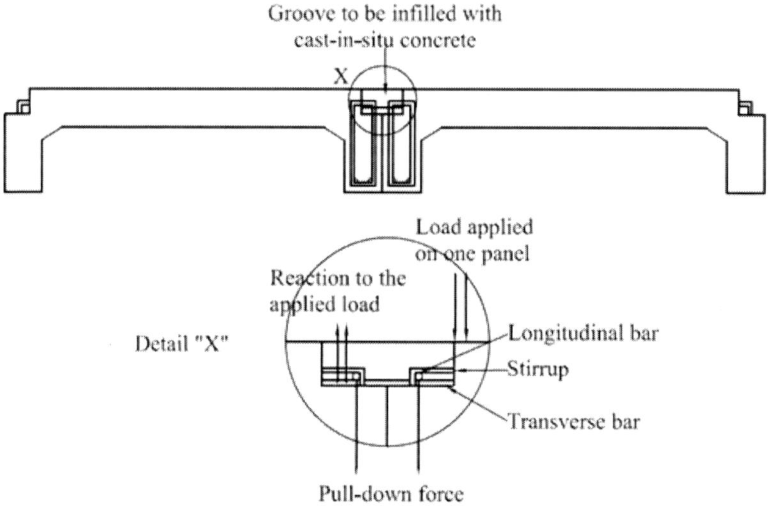

Figure 7.57. Load and reaction view at joint [2].

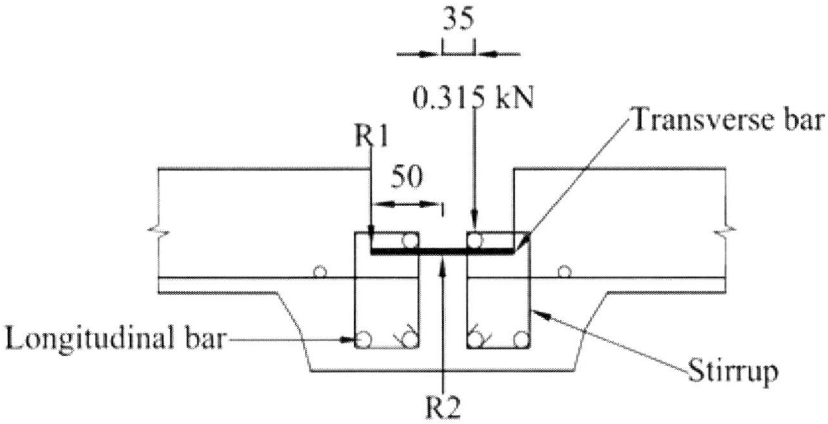

Figure 7.58. Sectional view of reaction diagram [2].

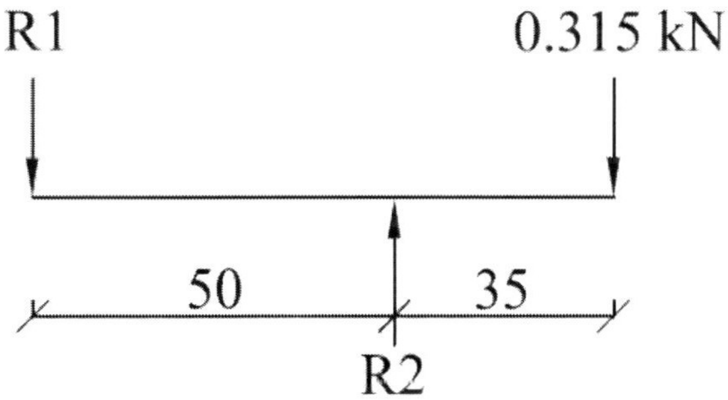

Figure 7.59. Free-body diagram [2].

Solving the equilibrium of moment,

$$\sum MR_0 = 0$$

$$(0.315x0.085)-(R_2x0.05) = 0$$

$$R_2 = 0.536KN$$

$$R_1 = 0.221KN$$

Each rod supports R_2= 0.536 kN and the total support reaction by four pairs = 0.536 x 4 x 2 = 4.29 kN >2.52 kN of 0.84m^2 area (OK)

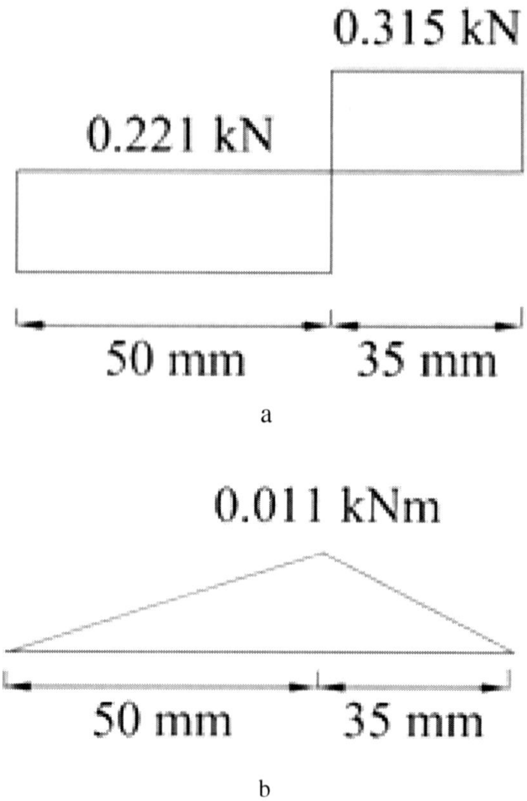

Figure 7.60. (a) Shear force diagram (b) Bending moment diagram [2].

Maximum bending moment= 0.035m×0.315kN= 0.011 kNm

Moment of inertia of 10mm steel rod, $I = \frac{1}{4}\pi r^4$

$$= \frac{1}{4}\pi(5)^4$$

I_{10}= 490.87 mm^4

For 10mm steel rod, $S_x = \frac{I}{c}$

$$= \frac{490.87}{5}$$

$$= 98.17 \text{ mm}^3$$

Allowable bending stress $= 0.66\, f_y$

$$= 0.66 \times 250$$

$$= 165 \text{ N/mm}^2$$

$$= 165 \text{ MN/m}^2$$

Developed bending stress $= \frac{M}{S_x}$

$$= \frac{0.011 \times 10^6}{98.17}$$

$= 112.05$ N/mm² < 165 N/mm² (OK)

Allowable Shearing Stress $= 0.4 f_y$

$= 0.4 \times 250$

$= 100$ N/mm²

Developed shear stress,

$$= \frac{0.315 \times 10^3}{\frac{\pi}{4}(10)^2}$$

$$= 4.01 \text{N/mm}^2 < 100 \text{N/mm}^2 \text{(OK)}$$

Allowable tensile stress $= 155$ MN/m²

Tensile stress = shear force /area $= \frac{0.315 x 10^3}{\frac{\pi}{4}(10)^2}$

$= 4.01$ N/mm² < 155 N/mm² (OK)

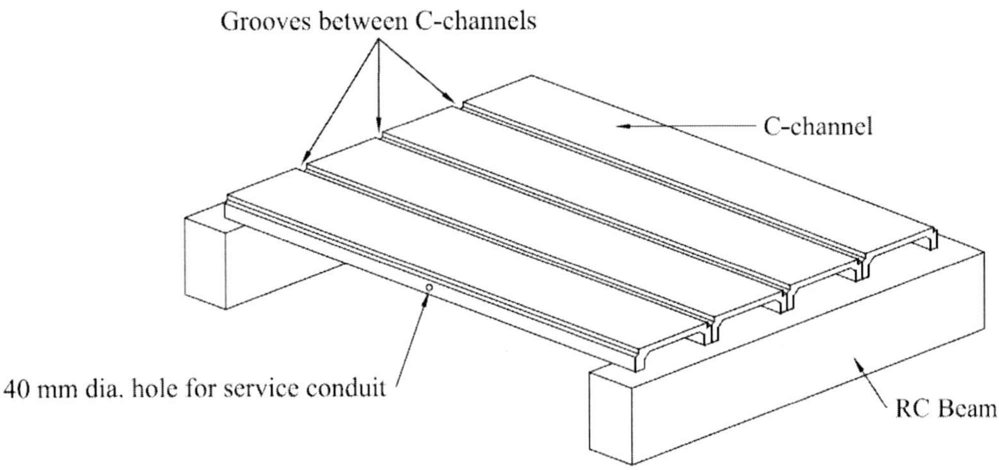

Figure 7.61. Conceptual monolithic floor formation of 5m span using precast C-channels made with OPS hybrid LWC [2].

Figure 7.62. Actual floor assembled with C-channels for forming monolithic floor [2].

7.5.14. Theoretical and Experimental Deflections of C-Channel with UDL

The deflection of the C-channel was calculated in the following way. The typical deflection is shown in Table 7.27 using a C-channel made with normal weight concrete.

Properties of C-channel:

$$E_s = 2x10^5 N/mm^2$$

$$E_c = 2.48x10^4 N/mm^2$$

$$d_1 = 240 - 25 - \frac{16}{2}$$

$$= 207\text{mm}$$

$$\text{Ratio, m} = \frac{E_s}{E_c}$$

$$= 8.06$$

$$\text{Neutral axis, } n = \frac{mA_s}{b}\left(\sqrt{1+\frac{2bd_1}{mA_s}}-1\right) \tag{7.15}$$

$$= \frac{(8.06)(804.2)}{1000}\left(\sqrt{1+\frac{2(1000)(207)}{8.06(804.2)}}-1\right)$$

= (6.48)(7.05)

= 45.7mm

$$\text{Cracking Moment of Inertia } I_c = \frac{bh^3}{3} + mA_s(d_1-n)^2 \tag{7.16}$$

$$I_c = \frac{(1000)(45.7)^3}{3} + (8.06)(804.2)(207-45.7)^2$$

$$= 2x10^8 mm^4$$

Total I_c for 3 units of C-channels as monolithic behaviour platform = $3\times2\times10^8$ mm^4

$$\text{Allowable deflection} = \frac{Span}{250} \tag{7.17}$$

$$= \frac{5000}{250}$$

= 20mm

Figure 7.63. Monolithic behaviour of platform using three precast C-channels of 5m span while loaded with middle C-channel [50].

Table 7.27. Experimental and theoretical deflections with uniform distributed loads onto C-channel of 5 m Span [50]

Layer of blocks	Total applied load, kN/m^2	Experimental deflection at different ribs of C-channel platform using LVDT, mm				Theoretical deflection of C-channel $=(\frac{5}{384}\frac{WL^4}{EI})$, mm	
		Rib-2	Rib-3	Rib-4	Rib-5	3 units of C-channels as monolithic platform, Ref. Figure 7.63	As one unit C-channel
1	2.09	0.5	0.5	0.5	0.4	1.14	3.42
2	4.18	0.7	0.8	0.8	0.8	2.28	6.85
3	6.27	1.2	1.2	1.2	1.2	3.42	10.28
4	8.36	1.8	2.0	2.1	2.0	4.57	13.71
5	10.45	2.7	2.8	3.0	2.8	5.71	17.13
6	12.54	3.5	3.8	3.8	3.6	6.85	20.56
7	14.63	4.4	4.5	4.5	4.5	7.99	23.99
8	16.72	5.3	5.3	5.1	5.2	9.14	27.42
9	18.81	6.0	6.2	6.0	6.0	10.28	30.84
10	20.90	7.0	7.2	7.3	7.0	11.42	34.27
11	22.99	7.8	8.0	7.9	7.8	12.56	37.70
12	25.08	8.7	9.0	9.0	8.8	13.71	41.13

7.5.15. Joint Integrity Test of C-Channel Platform Using UDL and Water Ponding

To minimize the shrinkage at the joints, a suitable concrete mix of equivalent structural grade was developed with heavy duty integral waterproofing admixture. The waterproof admixture is used especially for permanent waterproofing of mass and reinforced concrete structures, ready mixed and precast concrete etc. It does not contain calcium chloride and therefore, it will not cause corrosion to the reinforcements. This waterproofing improves the workability and plasticity of the mix, reduces water absorption and penetration, reduces the possibility of shrinkage and consequential cracking and crazing, and increases the compressive and tensile strength. The recommended dosage of this waterproof admixture is about 500 ml per 50 kg of cement. The compressive strength of the companion cubes of cast in-situ concrete used for infill was approximately 22.7 N/mm^2 at the age of 7-days.

Water ponding of about 50 mm deep, covering an area of 3.0 m long × 2.5 m wide, was formed on two inner C-channels as shown in Figures 7.64 to 7.66. Directly on one of these C-channels, UDL up to 11.57 kN/m^2 was applied using a series of concrete blocks (Table 7.28). Deflection at midspan was measured at every layer of blocks through two LVDTs placed at two points of adjacent rib soffits (Figure 7.67). At a UDL of 11.57 kN/m^2, the maximum recorded deflection for both adjacent ribs were 1mm, an indicator of monolithic joint of C-channels (Table 7.28 and Figure 7.68).

This load, 11.57 kN/m^2 was approximately 3.86 times higher than the design imposed load (DIL) of 3kN/m^2 (Table 7.24). Based on ACI Code, the immediate deflection due to design imposed load (DIL) for floors not supporting elements likely to be damaged should not exceed the allowable value of L/360 (i.e., 13.89 mm).

This also indicates that the groove dimensions provided on the C-channel were adequate and the inclusion and orientation of longitudinal and cross reinforcements underneath the protruding stirrups can effectively transfer the load to adjacent C-channels. It is also suggested that the cast in-situ concrete infill in the gap between C-channel ribs has played a significant role in shear resistance. The end connection of the C-channel with beam support was not considered in this investigation. It is expected that the C-channel floor structure with a proper connection both at the groove and end will achieve better structural behavior under real systems and applications.

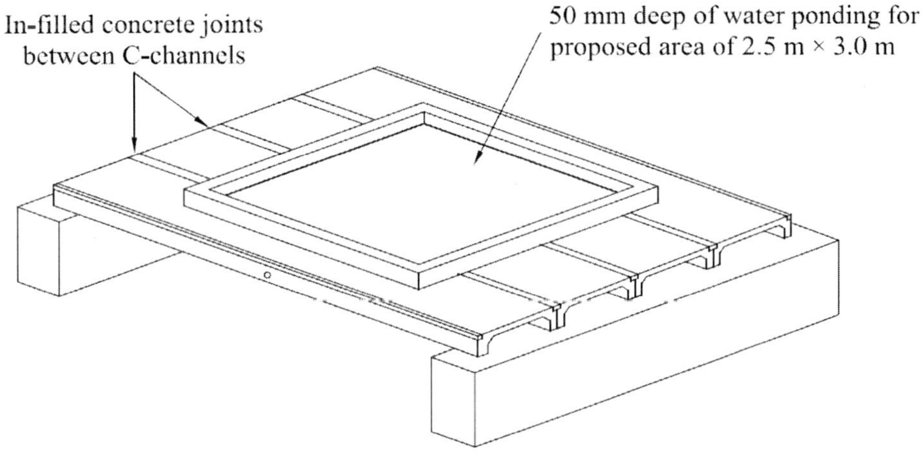

Figure 7.64. Water ponding of 50 mm deep for area of 3m long and 2.5m wide [2].

Figure 7.65. Water ponding of 50 mm deep for area of about 3m long and 2.5m wide [2].

Figure 7.66. Leak test on the joint line through water ponding of 50 mm deep with simultaneous UDL of 11.57 kN/m^2 placed onto one C-channel of 5 m span x 1 m wide area [2].

Figure 7.67. Perspective view of leak test on joint line through water ponding of 50 mm deep with simultaneous UDL of 11.57 kN/m^2 placed onto one panel of 5 m span x 1 m wide area [2].

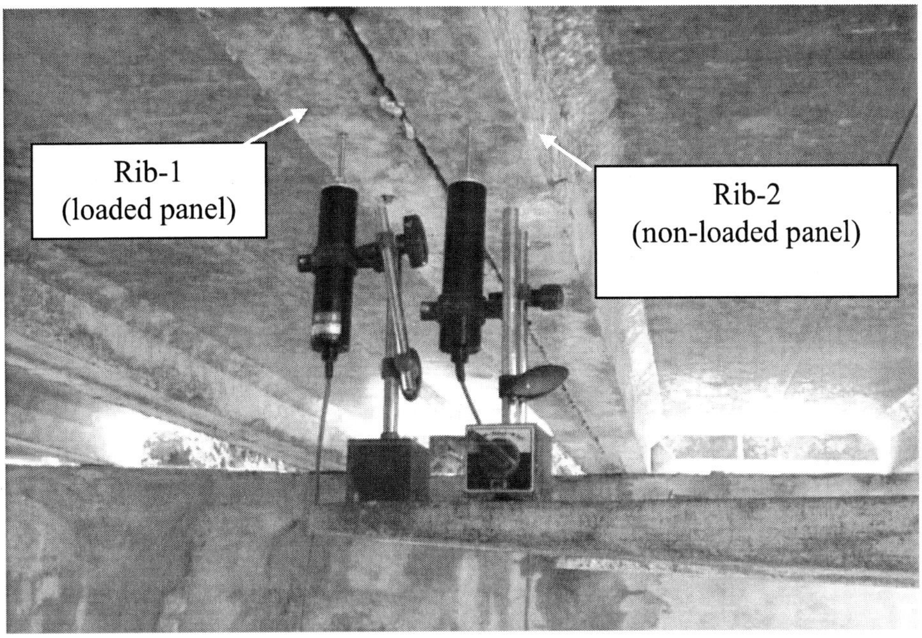

Figure 7.68. LVDTs set up at mid span for deflection check while leak test on joint line through water ponding of 50 mm deep with simultaneous UDL of 11.57 kN/m^2 placed onto one inner panel 5 m span x 1 m wide area, recorded deflections for both ribs with 1mm only [2].

Table 7.28. Referring to Figure 7.67, applied UDL and recorded deflections on C-channels made with OPS hybrid LWC [2]

Layer of blocks	Total of no. blocks	Cumulative load on area of 5m span and 1m wide (kN)	Applied UDL (kN/m^2)	Limit deflection, span÷360=5,000÷360=13.89mm Recorded deflection (mm)	
				Rib-1 (loaded panel)	Rib-2 (non-loaded panel)
1	65	7.23	1.45	-	-
2	65	14.46	2.89		
3	65	21.69	4.34		
4	65	28.92	5.78		
5	65	36.15	7.23		
6	65	43.38	8.68		
7	65	50.61	10.12		
Total	**455**	**57.84**	**11.57**	**1**	**1**

The conventional cast in-situ RC floor slab made with normal weight concrete generally appears monolithic with RC beams and is capable of transmitting lateral forces. The joint connection mechanism of C-channels as mentioned in Section 7.5.13 and with cast in-situ grouting in grooves proved the entire floor structure behaved monolithically. As such, under a gravity load (both permanent dead load and imposed load), the joint must transmit the load distribution across adjacent C-channels to the extent that the C-channel floor structure can deflect together. The load distribution depends on how well the connections of adjacent C-channels are made. This is why the lateral displacement of the units must be restricted by means of the reinforcement in the cross direction which gives the necessary force perpendicular to the joint when the adjacent elements tend to separate (Section 7.5.13). The required resistance of the cross reinforcement should be at least equal to the total vertical shear forces which have to be transmitted across the longitudinal joints. This then creates a pull-down mechanism, which pushes the adjacent C-channels together with the direct-loaded C-channel down effectively as shown in Figure 7.57. With this mechanism, it also provides an interlocking system to avoid independent displacement of C-channels. Thus this C-channel system does not need any topping concrete like other conventional precast flooring systems.

7.5.16. Joint Integrity Test of the C-Channel Platform with Water Ponding

A sign of water-leaking or dripping along the joint was closely and carefully examined for the duration of 120 hours to ensure leak proof cast in-situ concrete into groove. The intention of the water ponding test while a C-channel is loaded with UDL of 11.57kN/m^2 is to investigate the possibility of water leakage at the joint of C-channels.

Prior to water ponding, close visual inspection on the joint at the top and bottom faces was performed and there was no sign of any crack formation. Relative shrinkage differences between the C-channels and cast in-situ concrete in grooves can cause tensile stress along the interface, which will result in cracking and the leaking of water. Therefore, this visual examination was to ensure that the water would not leak through shrinkage cracks (if there were any) before being subjected to UDL.

Figures 7.65 and 7.66 show the water ponding on a platform made with C-channels covering the two intermediate C-channels with an approximate area of 3.0 m (long) x 2.5 m (wide) with roughly 50 mm of water. The UDL as shown in Figures 7.66 and 7.67 was applied directly onto one C-channel in which the first layer of blocks was under water. Throughout the test duration of 120 hours, no sign of water, leaking or dripping along the underside of the joint was observed. It further revealed that the adhesion between the cast in-situ concrete incorporating the use of waterproofing admixture and the precast C-channel was good. No leaking of water also indicated that there was no leaking crack or full-depth crack throughout the depth of cast in-situ concrete in the groove line. This shows that the joint as per design has not only effectively achieved the pull-down mechanism but also possessed excellent properties and integrity.

7.5.17. Provision of Service Conduits at the C-Channel Soffit

Service conduits are normally concealed for conventional RC floor systems. The special consideration for conventional precast floor systems is that it is always necessary to accommodate the openings for services in both directions. Usually, these openings are pre-mounted into the precast units prior to casting at the manufacturing plants. Other types of precast floors can easily accommodate the installation of external service conduits which can be done at the project site.

Owing to the cross-section of C-channels, the provision of service holes (Figure 7.54) and installation of external service conduits is addressed (Figures 7.69 and 7.70). The rib hole with a 40 mm diameter was less than one quarter of the overall rib depth. In general, the hole diameter should not exceed 25% of the relevant section i.e., h/4 for horizontal holes, where h is the overall depth [64]. The pipe was located at the midspan of the C-channel to avoid areas of high shear [65]. The openings should be centered slightly below the neutral axis to avoid high bending stress, and the corners of the openings must be circular rather than rectangular so as to avoid stress concentration [65]. No significant effect on the flexural behavior was noticed on the C-channel having such a 40mm diameter hole.

PVC pipe was used as the service conduit at the soffit of the C-channel (Figure 7.70). External service conduits can also be easily installed onto the underside of the C-channel. For precast floor units, it is very important that such fixings should be arranged in a manner not to damage or impair the reinforcements [64]. Since the flange thickness of the C-channel is 75 mm, it is not possible to embed the service conduits in the floor. [66] has specified that water pipes or electrical conduits embedded in the floor should have at least 38 mm of concrete cover on both the top and bottom. The anchorage zone for service conduits has been identified to be at the underside of the intersection between the rib and the flange (Figure 7.69). It is prohibited to install anchors near points of reinforcement because of safety interests [67]. With this aspect, the installation of such conduits in C-channels is feasible and very much comparable to other precast floor systems such as double tee flooring systems [61].

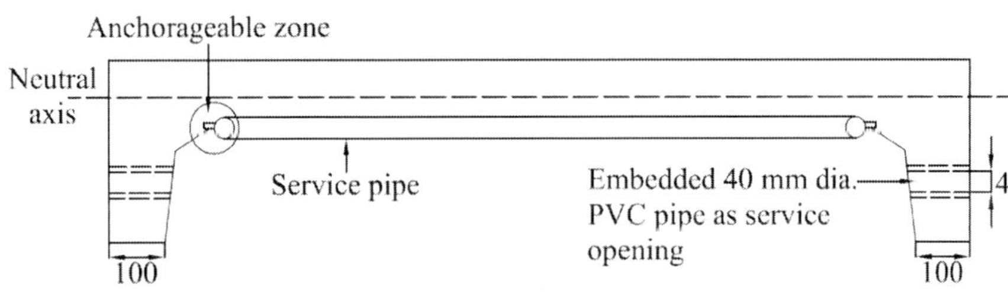

Figure 7.69. Location for service conduits in C-channels [2].

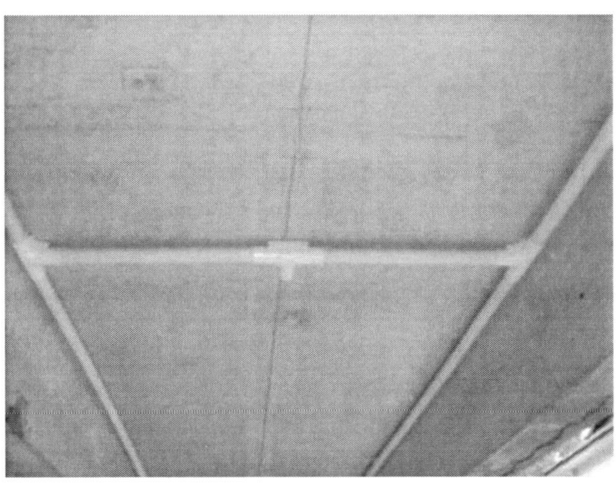

Figure 7.70. Reactions and forces induced through the bars [2].

7.5.18. Flange Load Capacity of C-Channels

The partition wall in a building is built at any location on the floor. It is required to determine the safe flange load capacity if any permanent line load is applied to the flange. Hence, the C-channel made with OPS hybrid LWC was tested under one- line load parallel to ribs to determine its flange load capacity as shown in Figure 7.71. The single-line load test was performed using an actuator of 1,000 kN capacity on 1-m long C-channel until failure. The effective span of the test was 900, as two ribs were placed onto supports.

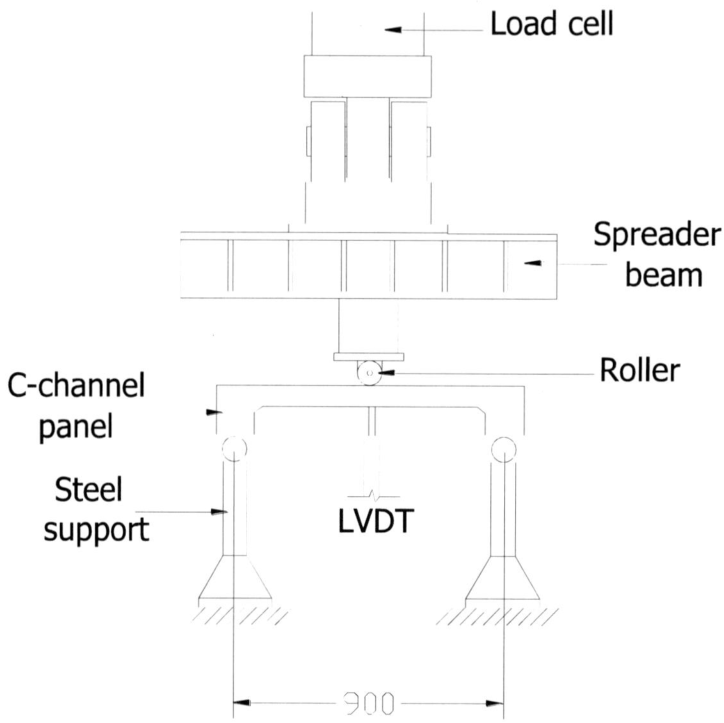

Figure 7.71. Test setup for flange load determination [2].

The experimental ultimate load of 14.74 kN produced one crack at the midspan. The experimental service load was found to be 8.67 kN, by dividing the experimental ultimate load by the factor of 1.7 [13]. At this service load of 8.67kN, the corresponding deflection was about 0.45 mm which was less than the limit value of 3.6mm (900/250 = 3.6 mm). Hence, the flange of C-channel met the serviceability limit state as per deflection. The developed crack width at service load was 0.06 mm, which was below the maximum limit of 0.3 mm as noted in design code. Again, this shows that the flange of the C-channel can meet the serviceability limit state as per crack width criteria. Figure 7.72 shows the tested C-channel after the flange load test.

Figure 7.72. Tested C-channel made with OPS hybrid LWC for flange load capacity [2].

7.5.19. Sustained Load of 3m High Brick Wall Built onto C-Channel

In real construction, usually the brick wall is built onto the floor. So, a brick wall of 3m high and 125mm thick including plaster as a non-load bearing wall was built along the 1,000 mm width of the C-channel as shown in Figure 7.73. The plastering is not seen in this figure. The deflection at midspan was measured with a special setting of linear ruler fixed on a stand for a period of 300 days.

The total dead load of the brick wall including the plastering on both sides was approximately 6.75 kN. This brick wall as a load test was done to verify the structural adequacy of C-channels during serviceability conditions for deflection and cracks.

Under a sustained brick wall load over 300 days, it showed the deflection of less than 2 mm, which was also less than the allowable deflection limit of 3.6 mm (900/250). The arch action developed in the brick wall reduces the bending moment in the supporting element like in this case, the C-channel. The arch action develops in the wall panel, which is supported on a beam, when the height/span ratio reaches the value of 0.6 [68]. In this study, the arch action took place as the ratio of 3.33. However, it can be concluded that the C-channel made with OPS hybrid LWC meets the serviceability limit state under prototype brick wall loading. Figures 7.74 and 7.75 show the typical floor and partition wall arrangement onto a C-channel floor system.

Figure 7.73. A 3m high unplastered brick wall built on C-channel [2].

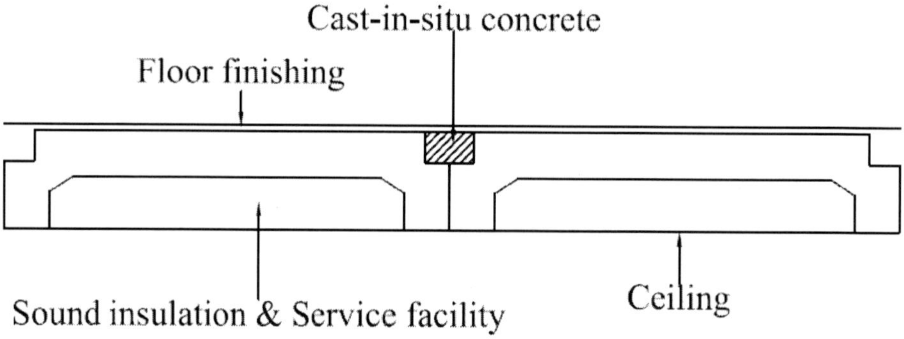

Figure 7.74. Typical final arrangement of C-channels in floor [2].

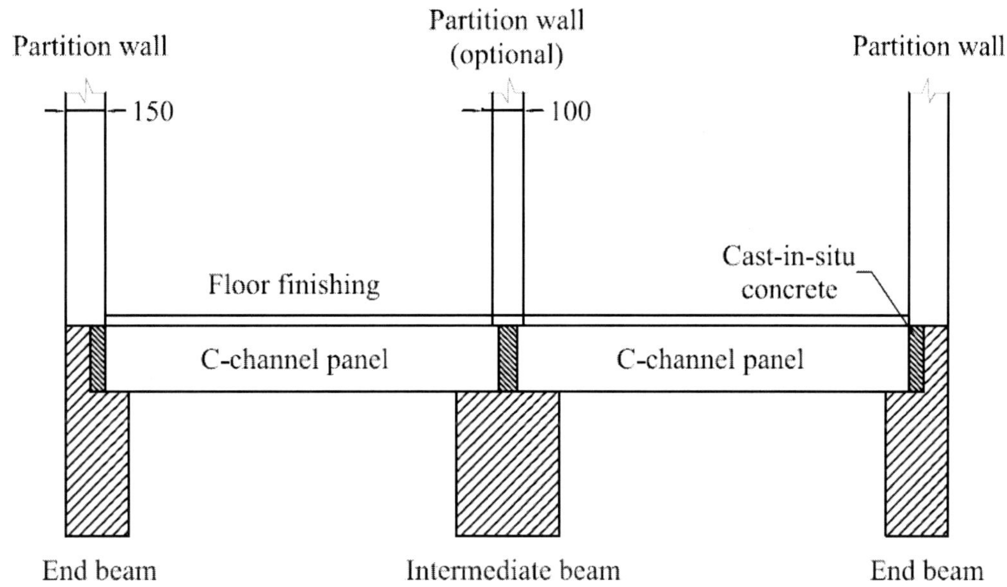

Figure 7.75. Typical partition wall set up onto C-channel floor [2].

7.5.20. Expressions for C-Channel Analysis Using BS and ACI Codes

(i) Ultimate Moment Calculation:

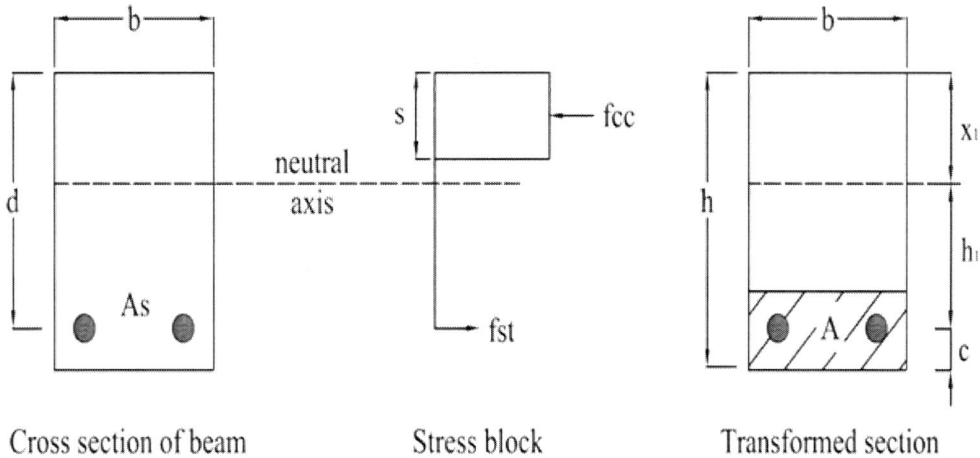

Figure 7.76. Beam cross section, stress block and transformed section [2].

As per BS 8110

$$M_{BS} = 0.95 f_y A_{st}\left(d - \frac{S}{2}\right) \tag{7.18}$$

where,

M_{BS} = ultimate moment based as per BS 8110, kNm

d = effective depth, mm

S = depth of equivalent rectangular stress block, mm

f_{cu} = characteristic concrete cube strength, N/mm^2

f_y = characteristic strength of reinforcement, N/mm^2

As per American Concrete Institute (ACI 318)

$$M_{ACI} = f_y A_{st}\left(d - \frac{S}{2}\right) \tag{7.19}$$

$$S = \frac{A_{st} f_y}{0.85 f_c' b} \tag{7.20}$$

where,

M_{ACI} = ultimate moment based as per ACI 318, kNm

f_c' = characteristic concrete cylinder strength (taken as 0.8 f_{cu})

(ii) Deflection Calculation

Deflection as per BS 8110

$$\Delta_{BS} = K l^2 \lambda \tag{7.21}$$

$$K = 0.125 - \left(\frac{a^2}{6}\right) \tag{7.22}$$

where,

K = constant that depends on the shape of the bending moment diagram

l = effective span of the member, mm

λ = curvature at mid-span, mm^{-1}

a = shear span, mm

Deflection as per ACI 318

$$\Delta_{ACI} = \frac{Pa\,(3l^2 - 4a^2)}{48E_cI_{eff}} \tag{7.23}$$

$$I_{eff} = \left(\frac{M_{cr}}{M_a}\right)I_g + \left[1 - \left(\frac{M_{cr}}{M_a}\right)^3\right]I_{crt} \le I_g \tag{7.24}$$

$$I_{crt} = \frac{bx^3}{3} + nA_s(d - x)^2 \tag{7.25}$$

where,

I_{eff} = effective moment of inertia of the section, mm^4

I_{crt} = cracked moment of inertia of the section, mm^4

Ig = gross moment of inertia of the section, mm^4

M_{cr} =cracked moment, Nmm

M_a = maximum unfactored moment in member at stage for which deflection is being computed, Nmm

b = width of member, mm

n = modular ratio

x = distance of the neutral axis from the extreme compression fibre, mm

(iii) Cracking Moment and Crack Width Calculations

Cracking moment

$$M_{cr(Theo)} = \frac{f_rI_g}{y_t} \tag{7.26}$$

where

$M_{cr(Theo)}$ = cracking moment, kNm

f_r = flexural strength of concrete, N/mm^2

I_g = gross moment of inertia of the section, mm^4

y_t = distance of the extreme tension fiber from the neutral axis, mm

(iv) Crack width Calculation as per ACI 318 [5]

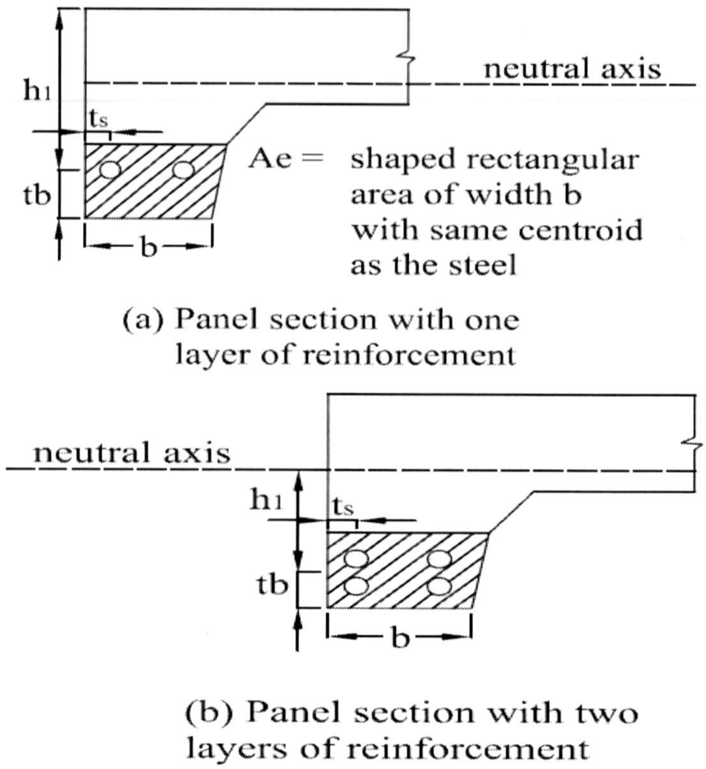

Figure 7.77. Cross section for crack width calculation [2].

The maximum crack width at the level of the reinforcement, w_{Theo}

$$W_{Theo} = \frac{1.1 \times \sqrt[3]{t_b A}}{1+\frac{2}{3}\left(t_s/h_1\right)} f_s \times 10^{-5} \tag{7.27}$$

where,

t_b = distance from the extreme tension fiber to the center of the adjacent bar, mm

t_s = distance from the side of the slab to the center of the adjacent bar, mm

A = average effective area of concrete in tension around each reinforcing bar (A_e/n, where A_e is the rectangular concrete area of width b and with same centroid as steel, b the beam width, and n the number of bars)

f_s = steel stress, N/mm^2

h_1 = distance from centroid of steel to neutral axis, mm

(v) End Rotation Calculation

End rotation of a simply-supported member, θ (rads)

$$\Theta_{Theo}\ (\text{rads}) = \frac{PxL^2(1-x)}{2EI} \tag{7.28}$$

where,

P = applied load, kN

x =distance of the loading point from the nearest end of panel, mm

L = span of the panel, mm

E = modulus of elasticity of concrete, kN/mm^2

I = second momend of inertia of the section, mm^4

7.5.21. Expression for Bearing Area Calculation of C-Channel

The effective bearing length l_b is taken as the least of (i) actual bearing length, (ii) one-half of (i) plus 100 mm, or (iii) 600 mm. Hence, the total effective bearing length is equal to the total rib width which is 100 + 100 = 200 mm (Figure 7.78).

The net bearing width l_w is equal to the nominal bearing width minus the spalling tolerances additive in the supporting and supported members (Tables 5.1 and 5.2) of [56] minus the allowances for constructional inaccuracies.

Load calculation for one 6m span C-channel

Weight of C-channel =2.83kN/m^2

Use 50mm Screed =0.05×1m×1m×24kN/m^3=1.2kN/m^2

Use Imposed load=3kN/m^2

Total UDL on panel =[1.4DL+1.6IL]6m span×1m width

=[1.4(2.83+1.2)+1.6(3)]6m^2

=62.65kN÷2 ends=31.33 kN/end

Each end has two ribs with 100mm width each

Use grade 25 concrete

Design ultimate bearing stress (Section 5.2.3.4) of [56])

Use dry bearing on concrete: $0.4f_{cu}$=0.4X25N/mm^2=10N/mm^2

Consider C-channel of 6 m span made with concrete compressive strength of 25 N/mm^2, the net bearing width (Section 5.2.3.2) of [56] =

Design ultimate support reaction per member

(effective bearing length × design ultimate bearing stress)

= 31.33×1,000N ÷ (200 mm × 10N/mm^2)

= 15.67 mm.

According to section 5.2.3.2 [56], the net bearing width should be at least 40 mm. Hence, 40 mm is considered. Considering the allowances for effects of spalling at supports [56]

(Table 5.1), the distance assumed ineffective for concrete below grade 30, plain or reinforced is 25 mm.

Referring to Figure 7.78, the minimum bearing width is 40 + 2(25) = 90 mm. Therefore, bearing width, l_w is taken as 100 mm.

Figure 7.79 shows heap of tested C-channels of 3 to 5m spans.

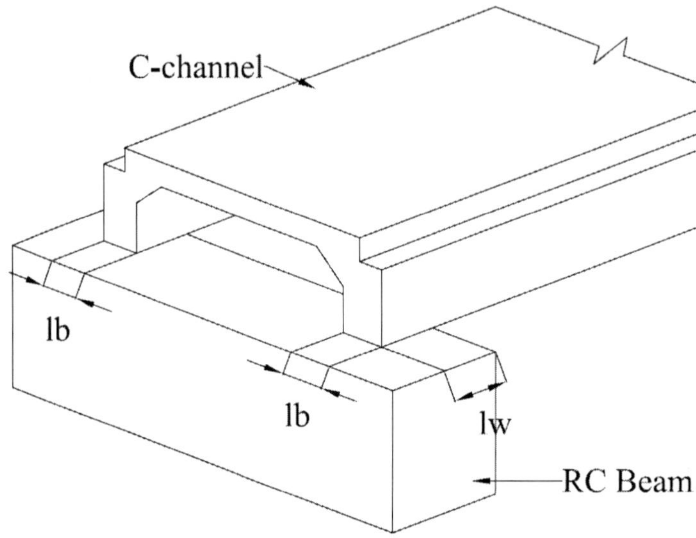

Figure 7.78. Bearing area of C-channel [2].

Figure 7.79. C-channels of 3 to 5m spans left after flexural investigation [2].

7.6. Chapter Remarks

It was observed that C-channels were more cost effective and efficient in terms of saving concrete, reduction of weight, higher load carrying capacity as compared with conventional full precast and semi-precast reinforced floor panels. The longer the span, the higher was the concrete savings. The cost reduction for C-channels was achieved through efficient material and optimum design. Therefore, the precast reinforced floor panel, C-channel made with OPS solid waste or conventional aggregate can be a promising alternative over the conventional concrete flooring systems.

REFERENCES

[1] ASTM C 494. *Standard Specification for Chemical Admixtures for Concrete.* Annual Book of ASTM Standards.

[2] Research Report. 2009. *Semi-precast lightweight flooring slab using OPS, CIDB Project no. LPIPM: CREAM/UPP04-02-10-04-11.* Part-1, Precast floor using OPS concrete by C. H. Ng. Universiti Malaysia Sabah.

[3] BS1881. Part 116. *Method for determination of compressive strength of concrete cubes.* British Standard Institution. London.

[4] BS 8110-2: 1997. *Structural Use of Concrete – Code of Practice for Special Circumstances.* London: British Standards Institution.

[5] P. Gergely, and L. A. Lutz. 1968. *Maximum Crack Width in Reinforced Flexural Members. Causes, Mechanism and Control of Cracking in Concrete.* ACI SP-20. American Concrete Institute, Detroit, pp. 87-117.

[6] J. H. Bungey, S. G. Millard and M. G. Grantham. 2006. *Testing of Concrete in Structures.* (4th edition). New York: Taylor & Francis.

[7] BS 6399. Part 1. 1996. *Loading for buildings. Code of practice for dead and imposed loads*, London, British Standards Institution.

[8] Building Department. 2004. *Code of Practice for Structural Use of Concrete* (Reprinted, incorporating Amendment No.1), The Government of the Hong Kong Special Administrative Region, Hong Kong.

[9] Anglo Technical Division. 2007. AA Specification 11410: *Design of Concrete Structures*, Issue 2, Anglo American.

[10] O. Mulia, S. P. Chiew and S. T. Lie. 2001. Moment Resistance of Steel I-Beam to CFT Column Connections with Composite Slab, Tubular Structures IX: *Proceedings of the Ninth International Symposium and Euroconference on Tabular Structures*, Düsseldorf, Germany, 3 -5 April 2001, pp 543 – 552.

[11] B. M. Ayyub, Al-Mutairi and P. Chang. 1996. Structural Strength of Bridge Decks Reinforced with Welded Wire Fabric, *Journal of Structural Engineering*, **122** (9): 989 – 997.

[12] R. I. Gilbert and Z. I. Sakka. 2007. Effect of Reinforcement Type on the Ductility of Suspended Reinforced Concrete Slabs, *Journal of Structural Engineering*, **133** (6): 834 – 843.

[13] M. A. Rashid and M. A. Mansur. 2005. Reinforced High Strength Concrete Beams in Flexure, *ACI Structural Journal.* **102**(3). pp. 462–471.

[14] R. I. Gilbert, Z. I. Sakka and M. Curry. 2006. Moment Redistribution in Reinforced Concrete Slabs containing Welded Wire Fabric, *The Tenth East Asia-Pacific Conference on Structural Engineering and Construction*, August 3-5 2006. Thailand, pp. 305 – 310.

[15] D. C. L. Teo, M. A. Mannan and V. J. Kurian. 2006. Structural concrete using Oil Palm Shell (OPS) as lightweight aggregate. *Turkish Journal of Engineering and Environmental Sciences.* **30**. pp 251-257.
[16] R. Park and W. L. Gamble. 2000. *Reinforced Concrete Slab*, 2nd ed., John Wiley & Sons, Inc. United States of America.
[17] ACI Committee 224. 1972. *Control of Cracking in Concrete Structures*, Proc. ACI. **69**. pp 717-753.
[18] C. K. Wang, C. G. Salmon and J. A. Pincheira. 2007. *Reinforced Concrete Design*, 7th ed., John Wiley & Sons, Inc. U.S.A.
[19] S. A. Ashour. 2000. Effect of Compressive Strength and Tensile Reinforcement Ratio on Flexural Behaviour of High-Strength Concrete Beams, *Engineering Structures.* **22**(5):413 – 423.
[20] T. U. Mohammed, H. Hamada and T. Yamaji. 2003. Relation between Strain on Surface and Strain over Embedded Steel Bars in ASR Affected Concrete Members, *Journal of Advanced Concrete Technology.* **1**(1). pp76 – 88.
[21] BS EN13747. 2005. *Precast concrete products – Floor plates for floor systems*, London: British Standard Institution.
[22] The Government of The Hong Kong Special Admistrative Region. *Code of Practice for Precast Concrete Construction*: 2003. Hong Kong.
[23] Halfen-Deha. 2003. *Precast lifting and fixing system*, Halften Limited,Germany.
[24] K. S. Elliott. 2006. *Precast concrete structures*, Elsevier Butterworth-Heinemann.
[25] Smorgon Steel Group. 2004. *Humeslab Technical Manual*, Australia.
[26] H. B. Basri, M. A. Mannan and M. F. M. Zain. 1999. Concrete using waste oil palm shells as aggregate. *Cement and Concrete Research.* **29**(4). pp 619-622.
[27] M. A. Mannan and C. Ganapathy. 2001. Long-term strengths of concrete with oil palm shell as coarse aggregate, *Cement and Concrete Research.* **31**(9). pp 1319-1321.
[28] H. B. Basri, M. A. Mannan, M. F. M. Zain and M. N. Islam. 2002. Effect of curing conditions on the properties of OPS-concrete, *Building and Environment.* **37**(11). pp 1197–1171.
[29] M. A. Mannan and C. Ganapathy. 2002. Engineering properties of concrete with oil palm shell as coarse aggregate, *Construction and Building Materials*, **16** (1). pp 29-34.
[30] M. A. Mannan and C. Ganapathy. 2004. Concrete from an agricultural waste-oil palm shell (OPS), *Building and Environment.* **39** (4). pp 411-418.
[31] M. A. Mannan, J. Alexander, C. Ganapathy and D. C. L. Teo. 2006. Quality improvement of oil palm shell (OPS) as coarse aggregate in lightweight concrete, *Building and Environment.* **41** (9). pp 1239-1242.
[32] D. C. L. Teo, M. A. Mannan and V. J. Kurian. 2005. Utilisation of solid waste Oil Palm Shell (OPS) in concrete production. *Proceedings of the International Conference on Natural Resources and Environmental Management*, Kuching, Sarawak, Malaysia.
[33] The Occupational Safety and Health Service. 2002. *Approved Code of Practice for The Safe Handling, Transportation and Erection of Precast Concrete.* Department of Labour, Wellington, New Zealand.
[34] Deha Transport Anchor System. 2003. *Technical Information KKT 03-E*, HALFEN-DEHA, Germany. 2003. Table 15.

[35] D. C. L. Teo, M. A. Mannan and V. J. Kurian. 2006. Structural Concrete Using Oil Palm Shell (OPS) as Lightweight Aggregate, *Turkish Journal of Engineering and Environmental Sciences.* **30**. pp 251-257.
[36] C. O. Orangun. 1967. The Bond Resistance between Steel and Lightweight-Aggregate (Lytag) Concrete, *Building Science.* **2**. pp 21-28.
[37] N. Chitharanjan, R. Sundararajan and P. D. Manoharan. 1988. Development of Aerocrete: A new lightweight high strength material. *The International Journal of Cement Composites and Lightweight Concrete.* **10.** pp 27-38.
[38] C. H. Ng, Z. Ideris, S. P. Narayanan, M. A. Mannan and V. J. Kurian. 2007. Engineering Properties of Oil Palm Shell (OPS) Hybrid Concrete for Lightweight Precast Floor Slab, Coventry University and UWM Centre for By-Products Utilization, *International Conference on Sustainable Construction Materials and Technologies.* UK.
[39] American Concrete Institute. 2002. *Building Code Requirements for Structural Concrete (ACI 318-02) and Commentary (ACI 318R-02).* American Concrete Institute. Farmington Hills, MI.
[40] C. H. Goodchild. 1997. *Economic Concrete Frame Elements.* Berks: British Cement Association.
[41] P. W. Matthew and D. H. Bennett. 1990. *Economic Long Span Concrete Floors.* Berks: British Cement Association.
[42] C. E. Reynolds and J. C. Steedman. 1992. *Examples of the Design of Reinforced Concrete Buildings to BS8110.* (4th edition). NewYork: Taylor & Francis.
[43] K. Baskaran and C. T. Morley. 2004. A New Approach to Testing Reinforced Concrete Flat Slab. *Magazine of Concrete Research.* **56**(6). pp 367 – 374.
[44] K. S. Elliot. 2006. *Precast concrete structures.* Elsevier Butterworth-Heinemann, Burlington.
[45] Code of practice for precast concrete construction. 2016. The Government of the Hong Kong Special Administrative Region. Hong Kong.
[46] C. H. Ng, N. S. V. K. Rao and M. A. Mannan. 2011. Precast C-channel floor for rural and estate housings, *Malaysian Construction Research Journal.* **9**(2). pp 63-72.
[47] D. C. L. Teo. 2007. A study on structural lightweight concrete using oil palm shell (OPS) aggregate, *PhD Thesis*, Universiti Malaysia Sabah, UMS, Kota Kinabalu, Sabah, Malaysia.
[48] ACI 318-05M. 2005. Metric Building Code Requirements for Structural Concrete & Commentary, *ACI Manual of Concrete Practice.* Detroit: American Concrete Institute.
[49] National Building Code: Part VI Structural Design, Section 7: Prefabrication and System Building, Bureau of Indian Standards.
[50] Research Report. 2009. *Semi-precast lightweight flooring slab using OPS, CIDB Project no. LPIPM: CREAM/UPP04-02-10-04-11.* Part-2, Precast floor using normal weight concrete by Doh Shu Ing. Universiti Malaysia Sabah.
[51] S. Mindess and J. F. Young. 1981. *Concrete.* New Jersey: Prentice Hall.
[52] B. Liu, Y. Xie and J. Li. 2005. Influence of Steam Curing on the Compressive Strength of Concrete Containing Supplementary Cementing Materials. *Cement and Concrete Research.* **35.** pp 994 – 998.
[53] J. Alexanderson. 1972. Strength losses in heat cured concrete, Swedish Cement and Concrete Research, Institute of Technology, Proc. No. 43 (Stockholm).

[54] T. K. Erdem, L. Turanli and T. Y. Erdogan. 2003. Setting Time: An Important Criterion to Determine the Delay Period Before Steam Curing of Concrete. *Cement and Concrete Research*. **33**. pp 741 – 745.

[55] A. M. Neville. 1999. *Properties of Concrete*. (4th edition). Malaysia: Longman.

[56] Erdogdu S. and Kurbetci S. 1998. Optimum heat treatment cycle for cements of different type and composition. *Cement Concrete Research*. **28.** pp 1595–1604.

[57] BS 8110-1: 1997. Code of practice for design and construction: British Standards Institution.

[58] Precast Flooring Federation. 2007. *Code of Practice: For the Safe Erection of Precast Concrete Flooring and Associated Components*. Leicester: Precast Flooring Federation.

[59] K. S. Elliott. 2002. *Precast Concrete Structures*. Oxford: Elsevier Butterworth-Heinemann.

[60] K. S. Elloitt. 2001. Precast Concrete Floors: The Future, Concrete, http://findarticles.com/p/articles/mi_qa5379/is_200110/ai_n21479794/. Retrieved 20 November 2009.

[61] K. S. Elliott. 1996. *Multi-storey Precast Concrete Framed Structures*. London: Blackwell Science Ltd.

[62] PCI Design Handbook. (5th edition). 1999. *Precast and Prestressed Concrete*. Chicago, IL, 1999.

[63] EN 1992-1-1. 2004. Eurocode 2: Design of Concrete Structures – Part 1-1: General Rules and Rules for Buildings. *European Committee for Standardization.*

[64] R. Bunn, D. Simpson and S. White. 2004. Services Integration with Concrete Buildings: Guidance for A Defect-Free Interface, BSRIA/The Concrete Society, Multiplex Medway Ltd.

[65] M. A. Mansur and K. H. Tan. 1999. *Concrete Beams with Openings: Analysis and Design*, United States of America: CRC Press LLC.

[66] ACI 302. 2004. Guide for Concrete Floor and Slab Construction (ACI 302.1R-04). *ACI Manual of Concrete Practice*. Detroit: American Concrete Institute.

[67] C. Lutz. 2002. Prestressed Hollow-core Concrete Slabs – Problems and Possibilities in Fastening Techniques. *Otto-Graf-Journal*. **13.** pp 53- 64.

[68] S. R. Davis. 1987. Composite Action. *Second International Seminar on Structural Masonry for Developing Countries*, March 23-26, 1987. Malaysia.

Chapter 8

LONG-TERM PERFORMANCES OF STRUCTURES BUILT USING OPS CONCRETE

Keywords: OPS concrete, lightweight concrete, Footbridge, Intermediate Terrance house, Sustained loading onto C-channel

8.1. INTRODUCTION

This chapter explains the very long-term performances of prototype structures such as footbridges and low cost houses, and prototype precast floor panels with a sustained load. All of these were made with OPS concrete either precast or cast in-situ. The footbridge was made with precast RC floor panels and topping concrete was done using plain OPS concrete as cast in-situ. It was built in 2001 and is still in service. The low cost house with the concept of an intermediate terrace unit was built in 2003. It is also still in service. The details of construction sequences and thermal performances of this house are explained. The prototype precast RC floor panel known as a C-channel was tested with a sustained load of more than 5 years.

8.2. 15 YEARS IN SERVICE OF FOOTBRIDGE BUILT USING OPS CONCRETE

An attempt was made to construct the prototype structure using OPS concrete to verify the laboratory findings of this new lightweight concrete. The footbridge was built using precast RC floor panels and topping concrete was made with plain OPS concrete as cast in-situ. The precast RC panel was 3m long. The 28-day compressive strength of the OPS concrete was 25 N/mm^2. It was built on the main campus of the Universiti Malaysia Sabah, Malaysia over a drain (Figure 8.1). The recorded relative humidity on this campus falls within 70-90% under a tropical environment having high rainfall. Based on published documentation, this footbridge is the longest serving prototype structure built using OPS concrete. Figures 8.1 to 8.5 show the different aspects of this footbridge.

Figure 8.1. Assembling of precast RC panels made with OPS concrete for footbridge construction in 2001 [1].

Figure 8.2. Footbridge built in 2001 using precast RC panel made with OPS concrete, still in service [courtesy, author].

Figure 8.3. Physical observation being conducted on footbridge as built using OPS concrete in a high humidity environment. This photo taken on 27 April 2012 [courtesy, author].

Figure 8.4. Colour paint done on footbridge soffit for monitoring surface conditions. This photo taken on 27 April 2012 [courtesy, author].

Figure 8.5. Exposed OPS coarse aggregate seen in deep brown colour (after spalled concrete from footbridge soffit during installation of hand railing). This photo taken on 27 April 2012 [courtesy, author].

8.3. 13 Years in Service of Low Cost House Built Using OPS Concrete

The purpose of the low-cost house built using OPS concrete was to promote OPS concrete in the construction industry. It was also intended to maintain thermal comfort naturally for such low-cost houses in the tropical environment. OPS solid waste is discharged at oil palm mills and is proven as a suitable construction material for the production of structural concrete grade. There are about 450 oil palm mills in Malaysia [2]. Oil palm mills run throughout the year and it is estimated that about 500 people from plantation workers to the mill manager are engaged at each mill. All mills are situated in rural and remote locations where the conventional construction materials, especially crushed stone as coarse aggregate are not available locally. OPS in such locations is the best option for the construction industry. Usually, at oil palm mills, the accommodations for staff are arranged as:

1) Manger Bungalow
2) Assistant Manager Bungalow
3) Semi Detached Artisan Quarters
4) Semi Detached Staff Quarters and
5) Terrace Type Labour Quarters

The low cost house of intermediate terrace was chosen as it has a natural air flow in one direction only. The size of this house is one story with a built up area of 58.68m^2 as shown in Figure 8.6. It has two bedrooms, one study, one living cum dinning room, one kitchen, one laundry, one shower and one toilet. The level of the laundry, shower and toilet being wet places was kept 50mm below the ground floor level. The prototype house is shown in Figures 8.7. About 17m^3 OPS concrete and 2,400 nos. 'OPS hollow blocks' were used in this house's construction. The details of the construction activities are shown below.

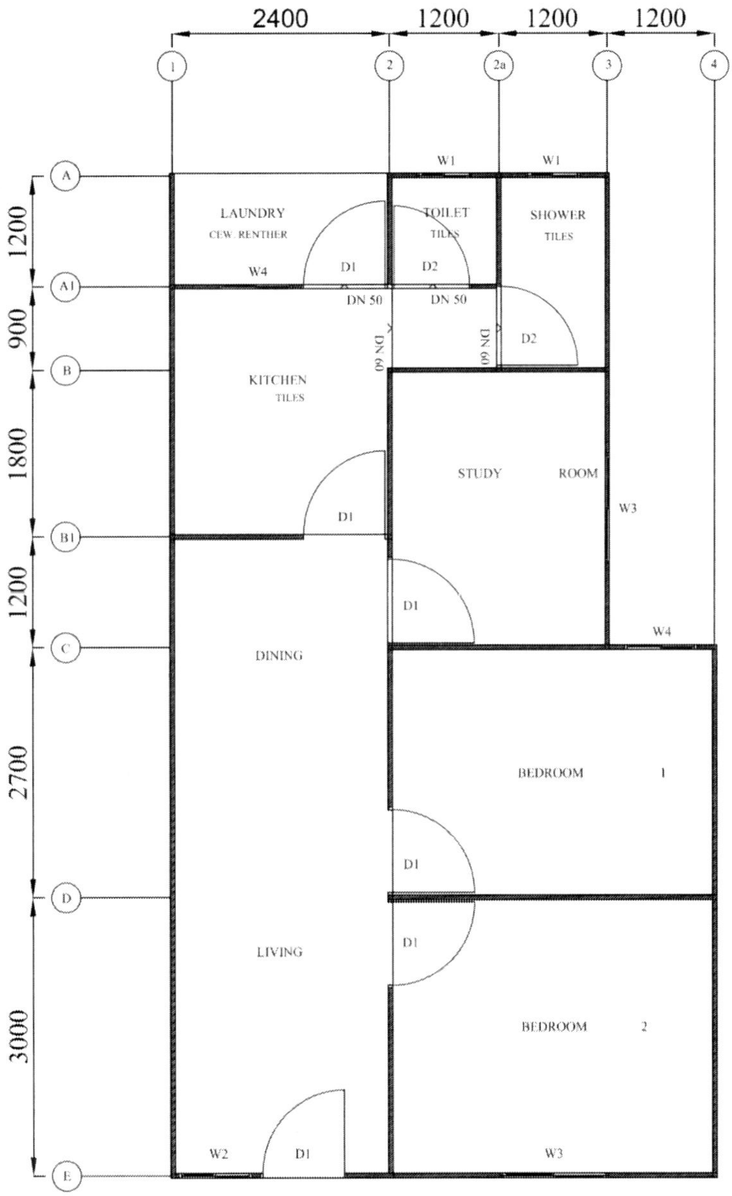

Figure 8.6. Floor plan of Low cost house built using OPS lightweight concrete [3].

Figure 8.7. 13 years in service with thermal comfort, this low cost house built using OPS concrete, was built in 2003 [courtesy, author].

8.3.1. OPS Hollow Block

It was an intention to produce wall blocks using locally available, alternative materials, considering economic, energy-efficient, and being user and environmental-friendly. The 'OPS hollow block' is made of OPS concrete in which OPS is the coarse aggregate along with cement, sand and water. The unit weight of this hollow block is about 1,200kg/m^3, whereas, conventional burnt clay brick has a unit weight about 1,800kg/m^3. So, OPS hollow block is about 33% less in unit weight. The properties of OPS hollow block are shown in Table 8.1. These hollow blocks used in house walls enhance more thermal comfort compared to conventional burnt clay brick. The OPS hollow block is 425mm long×125mm wide×190mm high including four holes. Figures 8.8 and 8.9 show the dimensions and full size of OPS hollow block. More than 3,600 OPS hollow blocks were produced at the Civil Engineering Laboratory, Universiti Malaysia Sabah in 2002. The OPS hollow block was an award winning product at the 32nd International Exhibition of Inventions, New Techniques and Products in Geneva, Switzerland, 31st March-4th April 2004.

Table 8.1. Properties of OPS Hollow Block [3]

Property	Value
Dimension of OPS Hollow Block, mm	420(L)x125(W)x190(H)
Compressive strength, minimum 2.8 N/mm^2	4.57
Drying shrinkage, maximum 0.06%	0.02
Void, minimum 25%	32
Water absorption, % (oven dry)	10
Block unit weight, kg	≈12

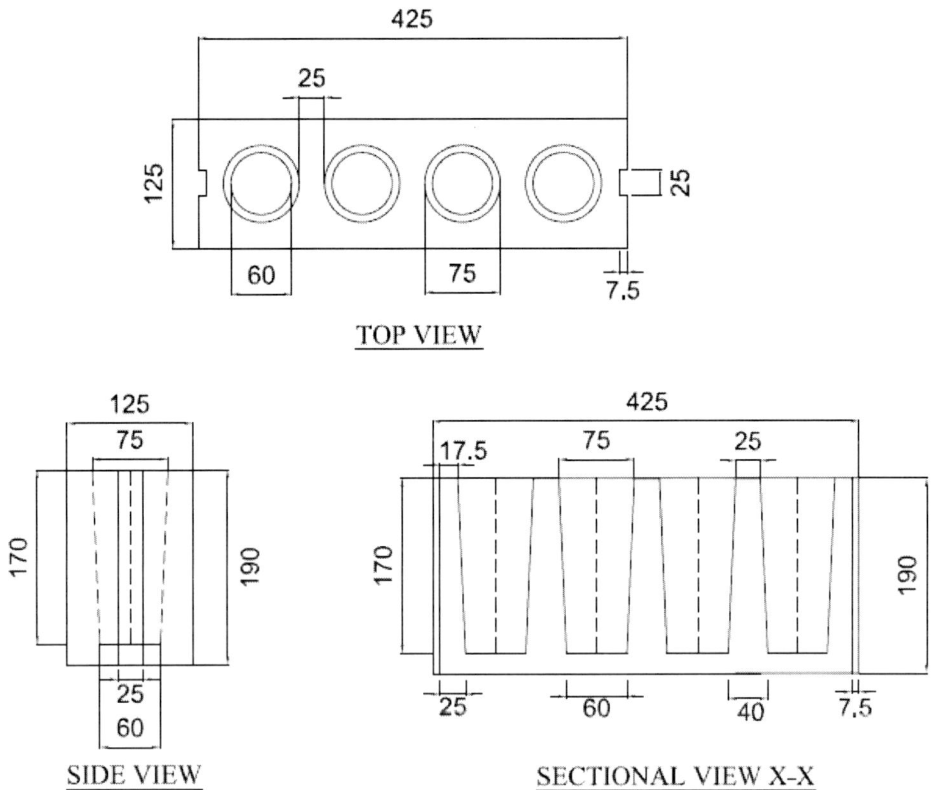

Figure 8.8. Dimension of OPS hollow block [3].

Figure 8.9. Full size of OPS Hollow Blocks made using OPS concrete, OPS is seen on top with dark brown colour [3].

8.3.2. Mixing of OPS Concrete at Site

The mixing of concrete for this project was done by hand at the site as shown in Figures 8.10 to 8.12. All materials prior to casting were kept at site. The footpath was used for casting this concrete manually. The river sand was placed onto the footpath, followed by cement and OPS. The OPS used was in saturated surface dry (SSD) condition. The required amount of water was used in mixing. The optimum mix in weight basis determined at laboratory was converted to volume basis to fit the site condition. Table 8.2 shows the mix proportions. The 28-day compressive strength obtained based on mixing at the site was in the range from 20 to 23 N/mm^2. The wheel barrow was used to measure the volume of OPS and river sand. The pail was used to measure the required water. The cement was used by the bag. The fresh concrete (Figure 8.12) was used to cast the ground beam, slab, lintel and roof capping beam immediately to avoid a segregation effect. It was manually compacted thoroughly. The concrete was placed in the formwork in layers not exceeding 150mm deep and each layer was compacted thoroughly. Tables 8.3 and 8.4 show the usage of OPS concrete and reinforcement details for different structural components.

Table 8.2. Mix content of OPS concrete for 28-day compressive St. of 20-23 N/mm^2 [3]

Materials	By weight (kg)	By volume		This concrete (%)	Ingredient ranges of normal weight concrete in common practice
		m^3	litre		
OPC (ASTM, Type 1)	50.0 (1 bag)	0.0354	35.4	16.97	9-15%
River Sand	88.5	0.0708	70.8	33.94	25-35%
OPS as coarse aggregate	47.1	0.0812	81.2	38.92	30-45%
Water	21.2	0.0212	21.2	10.16	7-15%

Figure 8.10. Oil palm shell (OPS), river sand, OPC and OPS hollow blocks kept ready at construction site [3].

Figure 8.11. Placement of oil palm shell (OPS) with SSD condition over sand and cement layers manually [3].

Figure 8.12. Manually mixing of OPS concrete at the open footpath [3].

Table 8.3. Use of OPS concrete in building components [3]

SSl	Materials	Usage area
1	OPS concrete, 28-day compressive St. of 20-23 N/mm^2, Total concrete used $=17m^3$	Use as reinforced concrete: • Ground beam of 225mm (W)X450mm (T) • Apron slab of 100mm thick • Ground slab of 100mm thick • Lintel of 125mm(W)X150mm (T) • Wall capping of 125mm(W)X150mm (T) • Roof beam of 190mm (W)X200mm (T) As unreinforced concrete for: • Lean concrete of 50mm thick
2	OPS hollow blocks using OPS concrete: Hollow block size = 425mm (L)X125mm (W)X190mm(T) Void/block=32% Water absorption ≈ 10% Weight/block ≈ 12 kg Crushing load/block ≈ 270kN Total hollow blocks used =2,400 nos.	Wall and Ventilators

Table 8.4. Reinforcement details of structural components [3]

Sl	Structural component	Size	Main reinforcement	Stirrup
1	Ground beam	225mm (W)X450mm (T)	4T10 as nominal	R8-250 as nominal
2	Ground Slab	100mm thick	Wire mesh of A6 as nominal	-
3	Lintel	125mm(W)X150mm(T)	4R8 as nominal	R6-200
4	Wall capping beam	125mm(W)X150mm(T)	4T10	R6-200 on wall and R6-100 on overhang part
5	Roof beam	190mm(W)X200mm(T)	4T10	R6-100

8.3.3. Ground Beam

The house was constructed on recently filled soil on the flat ground as shown in Figure 8.13. The foundation trenches were manually excavated with 275mm wide and 300mm deep to facilitate the concrete casting work (Figure 8.14). The trench was compacted manually for obtaining a uniform level of trench. After casting the ground beam, the excavated soil was used for backfilling the trench.

The formwork in this work was made using solid timber planks at both vertical sides of the trench (Figure 8.15). No plank was used at bottom of trench. The formwork was done with adequate bracing or being tied and capable of withstanding the pressure resulting from the fresh concrete used during compaction. It was also done to ensure sufficient tightness to hold the fresh concrete in the required shape without leaking and also maintaining the alignment of the ground beam. The reinforcement cages for ground beams were fabricated at the site and were placed in the formworks with plastic sheets as a protector during the casting of concrete. The nominal reinforcement of the ground beam was 2T10 at the bottom, 2T10 at the top and stirrups with R8 at 250mm intervals. The reinforcement details for ground beam, ground slab and apron slab are shown in Figures 8.16 and 8.17. The placement of plastic sheets and reinforcement cages for ground beams are shown in Figure 8.18. The partly filled ground beam with OPS concrete is shown in Figure 8.19. The reinforcement cages were held firm in the formwork against displacement. The concrete cover blocks were used. The formwork after casting the ground beam was struck until concrete had obtained the adequate compressive strength. All formworks were removed without shock or vibration to the concrete. The internal beam formwork was removed after 2 days of casting for facilitating ground slab preparation. The external beam formwork was removed 7 days after ground slab casting. The non-destructive test, NDT was done to check the compressive strength of OPS concrete upon removal of formwork from ground beams (Figure 8.20). The backfilling and compaction were done using excavated soil and river sand before the ground slab casting.

Figure 8.13. Proposed site with flat ground of less than 5 years in fill soil [3].

Figure 8.14. Manually excavated shallow foundation trench of 275mm wide and 300mm deep for preparation of ground beam [3].

Figure 8.15. Wooden formwork installation for ground beam to be made using cast in-situ OPS concrete [3].

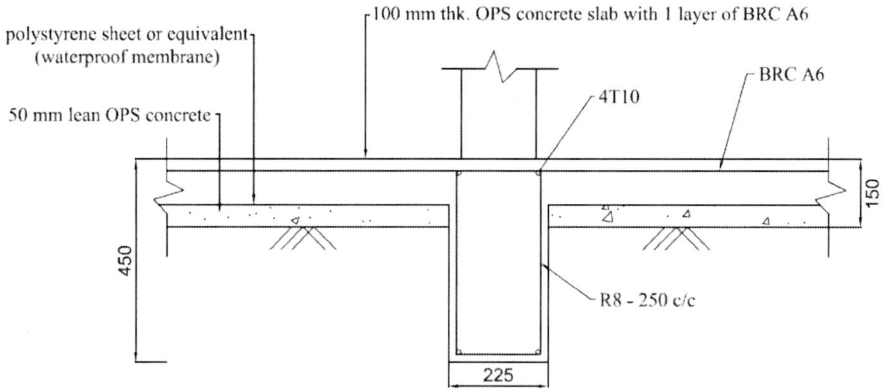

Figure 8.16. Reinforcement details of internal ground beams and ground slab to be cast in-situ using OPS concrete, lean concrete to be made also using OPS concrete [3].

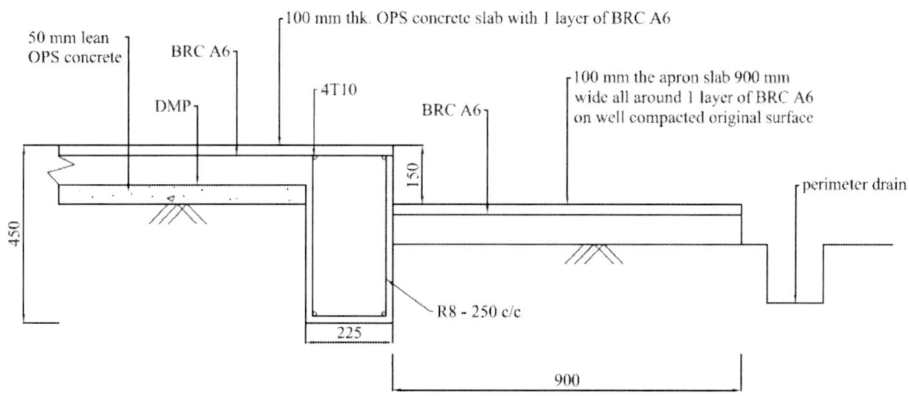

Figure 8.17. Reinforcement details of external ground beam and ground slab to be cast in-situ using OPS concrete, Apron slab to be made also using OPS concrete [3].

Figure 8.18. Set up of plastic sheets and reinforcement cages prior to the casting in-situ of ground beams [3].

Figure 8.19. Ground beams partly filled with fresh OPS concrete [3].

Figure 8.20. Non-destructive testing using rebound hammer to check concrete compressive strength upon removal of formwork [3].

8.3.4. Ground Slab

The floor area after ground beam was backfilled with river sand and levelled manually (Figure 8.21). OPS concrete shown in Figure 8.12 was used for 50mm thick lean concrete and levelled (Figure 8.22). The plastic sheet was placed permanently onto the lean concrete immediately after casting to stop evaporation, crack formation and stop rising moisture from ground. A few days after the casting of lean concrete, the nominal reinforcement using wire mesh of A6 (200mm spacing at both sides with 6mm round rod) with required cover blocks was placed (Figure 8.23) for ground slab casting. The overlapping of wire mesh was also provided as per specification. The ground slab was cast using OPS concrete as used for ground beam and lean concrete (Figure 8.24). Figures 8.12 and 8.24 show the fresh OPS concrete in very homogeneous condition. The ground slab was compacted and levelled manually (Figure 8.25). The thickness of the ground slab was 100mm. The plastic sheet was placed immediately after the levelling of the ground slab to prevent water loss from direct sun (Figure 8.26). The entire slab was cast in three stages within a week. The old concrete surface was roughened before placement of fresh OPS concrete for better bonding. Both the shower and toilet area being lower by 50mm were cast at a later day (Figure 8.26).

Figure 8.21. Manual compaction of backfill with river sand for the ground floor [3].

Figure 8.22. Formation of 50mm thick lean concrete using OPS concrete onto river sand backfill [3].

Figure 8.23. Placement of plastic sheet underneath the wire mesh, A6 with 200mm spacing at both sides with 6mm dia. round rods [3].

Figure 8.24. Placement of fresh OPS concrete for ground slab casting [3].

Figure 8.25. Manual levelling of ground slab immediately after placement of fresh OPS concrete [3].

Figure 8.26. Covering of ground slab with plastic sheets immediately after levelling the fresh OPS concrete, both shower and toilet area left without casting [3].

Figure 8.27. OPS Hollow Blocks ready for wall construction [3].

8.3.5. Load Bearing Walls Using OPS Hollow Blocks

OPS hollow blocks were used to make load bearing walls (Figures 8.27 and 8.28). No reinforced concrete (RC) columns were used in this house. The hollow blocks were arranged in layers at staggered form to avoid vertical crack formations in the walls. The required quantity of mortar for each batch with one part cement (ASTM Type I) and three parts river sand including w/c ratio of 0.4 were mixed manually for one hour. The practice of the local construction industry on mortar mix proportions was followed as river sand was too fine having fineness modulus (FM) of less than 1.2. No wire mesh was used in the block layers for this wall construction. The local construction industry uses wire mesh at every fourth layer of brick whether made with clay or sand-cement. The horizontal and vertical alignments of layers were made using tread and wood planks respectively (Figures 8.28 and 8.29). Eleven and ten layers of hollow blocks were used below and above the lintel respectively to suit the need (Figures 8.30 and 8.31). The hollow face of OPS hollow blocks were placed onto thin bedding mortar to form partly hollow walls for better thermal comfort in the house. Normally, house walls receive the heat from solar radiation in the morning and afternoon.

Figure 8.28. Making of load bearing walls using OPS hollow blocks manually without any wire mesh in the block layers. The horizontal alignment was done using thread [3].

8.3.6. Lintel and Capping Beam

The solid wood formwork was used for casting the lintel. The lintel was constructed at the top level of the door and window for the entire load bearing wall. This has the advantage of ensuring uniform stress distribution from the upper part of wall. Usually, such lintel placement does not allow any crack propagation either from down to top or top to down in the wall. The lintel size was 125mm wide and 150mm thick. The nominal reinforcement with longitudinal bars of 4R8 and stirrups of R6 at 200mm spacing was used. Figure 8.32 shows the wooden formwork arrangement prior to lintel casting using OPS concrete. Figure 8.33 shows the pouring of OPS concrete into formwork for lintel casting. The fresh concrete was compacted carefully as lateral support for formwork and was done using timber planks nailed at wall as shown in Figure 8.32. Leak-proofing was done before pouring of fresh OPS concrete.

Figure 8.29. Work progress on the load bearing wall using OPS hollow blocks, vertical alignment at corner done using wood planks [3].

Figure 8.30. Full height of load bearing wall using OPS hollow blocks including lintel placed above door and window level, and the apron slab in preparation [3].

Figure 8.31. Another view of the full height of the load bearing wall using OPS hollow blocks including lintel placed above door and window level, apron slab in preparation [3].

Figure 8.32. Arrangement of wooden formwork for lintel casting, wooden door frame also placed below lintel [3].

The capping beams were used at two levels of the wall such as the lower and upper part of the wall (Figures 8.34 and 8.35). The capping beam size was similar to lintel. The reinforcement with longitudinal bars of 4T10 and stirrups of R6 at 200mm spacing was used for lower wall capping part and 4T10 and R6 at 100mm spacing for the overhang. The upper capping beam was made using 4T10 longitudinal bars and the stirrups with R6 at 100mm spacing. Figure 8.34 shows formwork preparation for capping beam at the lower level. OPS concrete made at site (Figure 8.12) was used for capping beam. Figure 8.36 shows the overhang capping beam for the lower roof.

Figure 8.33. Pouring of OPS concrete into formwork for lintel casting [3].

Figure 8.34. Preparation of wooden formwork for wall capping beams including the overhang [3].

Figure 8.35. Wall capping beam including overhang made at lower level and upper level of load bearing wall using OPS concrete [3].

Figure 8.36. Overhang wall capping beam for lower level of wall using OPS concrete [3].

8.3.7. Roof, Ceiling and Passive Cooling System

Figures 8.37 to 8.43 except Figure 8.39 show the details of roof, ceiling and natural ventilation system. The pitch of the roof was made 15^0 and 17^0 to adjust the plan (Figure 8.38). The purlins were fixed on the wall capping beams using wall plugs. Galvanised self-drilling screws complete with neoprene washers were used in fastening the cladding to the purlins. The holes for fasteners were drilled at the centreline of the corrugation directly over the supporting purlins.

The mix used in mortar was also used in plaster work for both internal and external wall surfaces. The plastering as wet work inside the house was completed before installation of mechanical and electrical works, and ceilings. The suspension straps with solid timber of 50mm square were fixed with walls using suitable nails. The 3mm thick plywood as ceiling panels were nailed with suspension straps. Access openings from the ceiling into the attic area were provided. The light fittings, electric conduit and ceiling fan facilities were provided. The electricity was provided from the nearest power supply point.

The natural ventilation system above the ceiling was made using two layers of half cut OPS hollow blocks for the lower roof and three layers of half cut OPS hollow blocks for the upper roof. This house represents the intermediate unit of the terrace house, where natural air flow happens in one direction only. However, two layers of openings were provided (Figure 8.38). The outside air enters the narrow conduits (blocks act as narrow conduit) as compressed air resulting in cool air formation that reduces the attic temperature. Thus, the house can provide a passive cooling system naturally as the roof receives more heat from the sun through radiation.

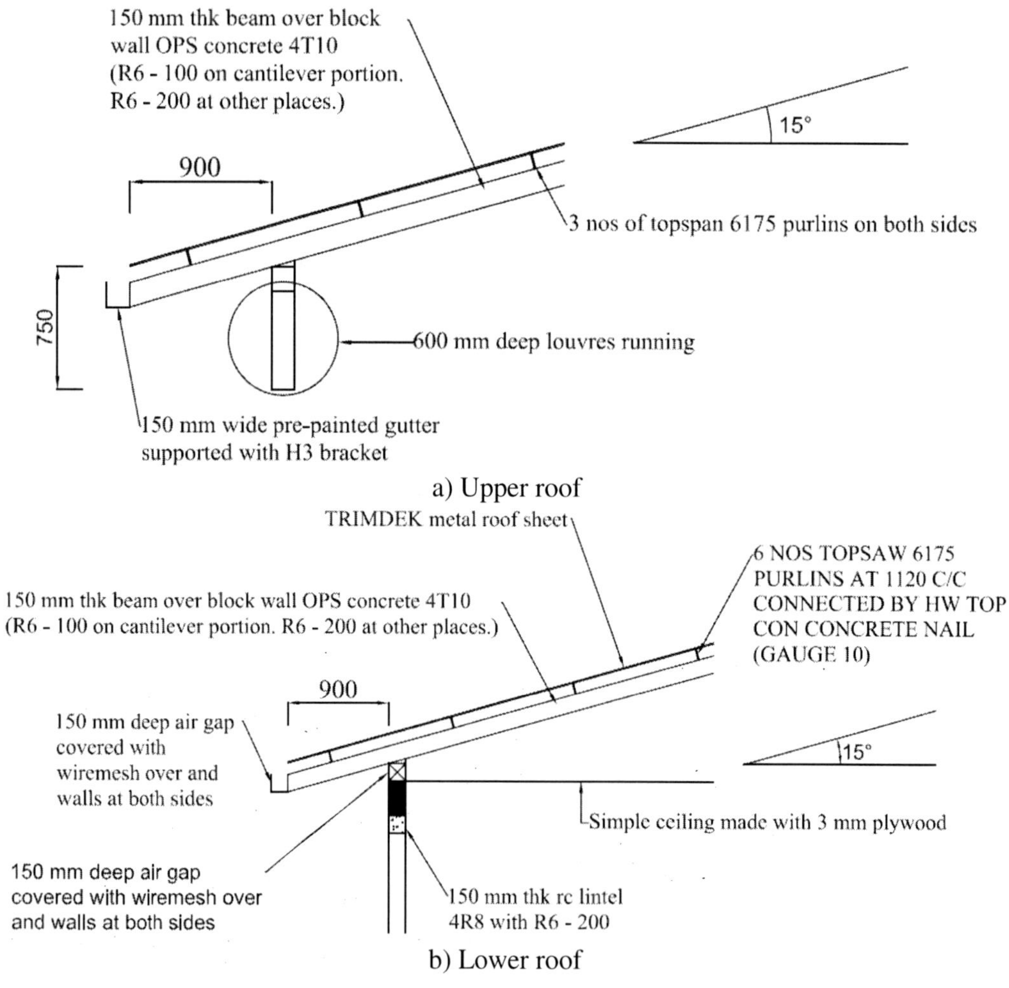

Figure 8.37. Roof details for both lower and upper roofs [3].

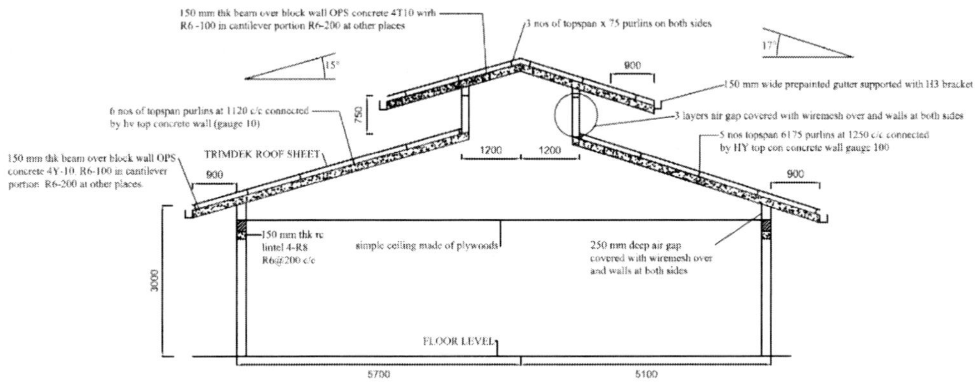

Figure 8.38. Longitudinal section of low cost house built using OPS concrete [3].

Figure 8.39. Plastering work on exterior surface in progress [3].

Figure 8.40. Plastering works for both internal and exterior surfaces of wall nearing finishes, seen house main entrance [3].

Figure 8.41. Completion of painting works on exterior and interior wall surfaces [3].

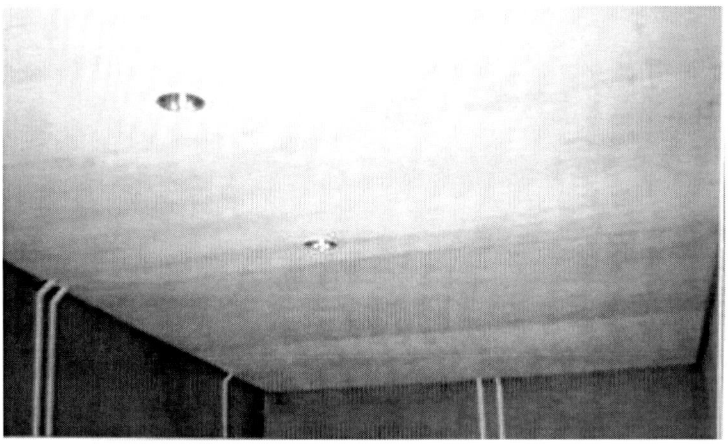

Figure 8.42. Fixation of 3mm thick plywood as ceiling, electric conduits and bulb holders at ceiling [3].

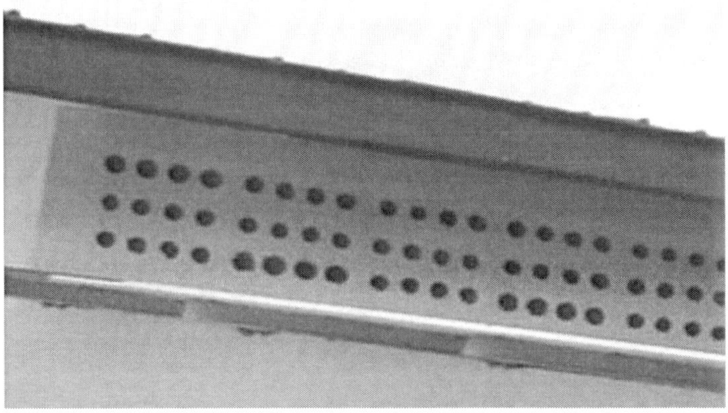

Figure 8.43. Provision of natural ventilation with mosquito net inside as passive cooling using half cut OPS hollow blocks [3].

8.3.8. Doors and Windows

Solid timber doors with decorations were used. The PVC doors were used for the bathroom and toilet. Windows were anodised aluminium frames with tinted glass. The windows for the bathroom and toilet were louver type. The details of doors and windows are shown in Figure 8.44.

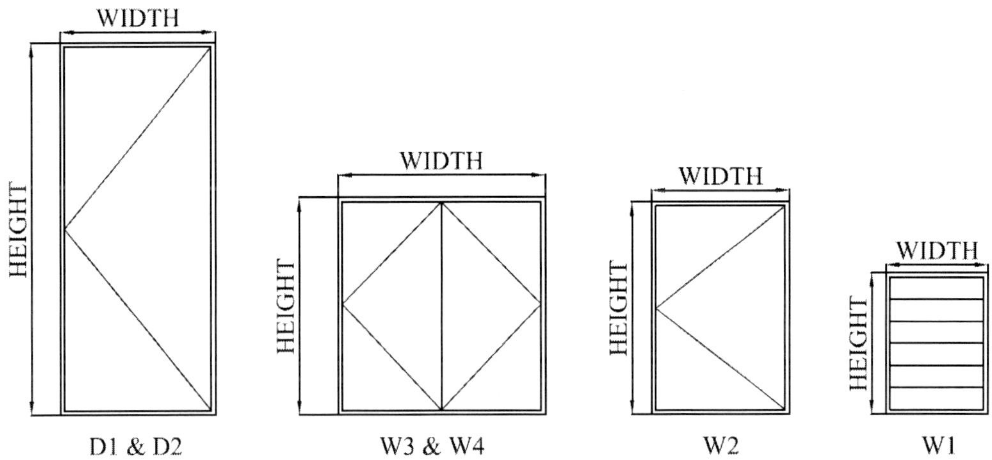

a) Door and Window types

DOOR SCHEDULE				
NO.	WIDTH	HEIGHT	DESCRIPTION	UNIT
D1	900	2100	SOLID TIMBER DOOR	6
D2	850	2100	PVC SLIDING DOOR	2

WINDOW SCHEDULE				
NO.	WIDTH	HEIGHT	DESCRIPTION	UNIT
W1	600	800	ALUM. LOUVRE WINDOW	2
W2	800	1200	ALUM. SINGLE CASEMENT WINDOW	1
W3	1200	1200	ALUM. TWIN CASEMENT WINDOW	2
W4	800	1200	ALUM. TWIN CASEMENT WINDOW	2

b) Dimension of doors and windows

Figure 8.44. Details of doors and windows used in this low cost house [3].

8.3.9. Thermal Comfort Study

Generally, the roof receives more solar radiation than walls during the peak outdoor temperatures. Thus, in the house it increases heat and enhances intolerable overheating during the daytime. The roof design needs particular attention as the main envelope of building, which is partly exposed to solar radiation. The thermal performance of a roof depends on many factors such as thermal insulation, ventilation, thermal mass etc. The conduction, convection and radiation transfer the heat through roof system. The roof is the surface of the building most affected by solar radiation, whereas air trapped between the ceiling and roof is considered as a major source of heat. The thermal conductivity of a metal roof is higher than concrete or clay tile. Convection is largely influenced by the direction of heat flow, shape and roughness of the solid surface, and type of fluid flow. The roof surface receives short wave radiation from the sun and long wave radiation from surrounding buildings and objects. The radiation in the attic is long wave radiation and can be reduced with a radiant barrier such as aluminium foil for its low thermal emissivity. The roof ventilation works to reduce the attic temperature, which reduces the heat transfer through the ceiling. The thermal comfort requirement needs to reduce the rate of heat gain from outside to inside. Excessive heat transfer (radiation) through the roof and walls causes thermal discomfort in a tropical zone. The passive cooling system in the building has great economic and environmental benefits. According to MS1525: 2007 [4], important factors for passive cooling design are:

1) Building configuration
2) Building facade
3) Building materials
4) Building orientation
5) Effective room depth
6) Fenestrations
7) Floor to ceiling depth
8) Internal layout
9) Landscaping and shading
10) Location of cores
11) Roof design and colour

Table 8.5. Materials used in walls, roofs, ceilings and attic conditions of three different houses [courtesy, author]

Description	Materials usage			Attic condition
	Wall	Roof	Ceiling	
House-1: Conventional one story house with built up area of 69.39m^2,	Bricks	Gable roof (metal roof and aluminum foil under roof sheet)	Gypsum board	Sealed
House-2: Conventional one story house with built up area of 79.17m^2	2mm UAC lightweight board made with treated cellulose fibre, OPC and refined sand	Gable roof (metal roof without insulation)	UAC board	Sealed
Low cost house built using OPS lightweight concrete with built up area of 58.68m^2	OPS hollow blocks	Gable roof (metal roof) with special design for thermal comfort	3mm thick Plywood	Well ventilated with 1/13 net free ventilation area

Thermal performances were studied considering this house and two other conventional one story houses built in the same city, Kota Kinabalu, Sabah, Malaysia. The materials used in the walls, roofs, ceilings, and attic condition are shown in Table 8.5. Table 8.6 shows the measured maximum and minimum temperatures during 28 days from 9 July 2004 to 5 August 2004. The HOBO thermometers were placed at center of the house, attic and outside to measure temperatures at one hour intervals. The doors and windows were kept closed during the data capturing time.

In comparison with other two conventional houses, the low cost house with two layers of ventilation including OPS hollow blocks in the walls has effectively shown the reduction of heat transmission. The thermal heat, caused by radiation absorbed by galvanized metal sheets, penetrates into the attic and passes to the interior. Convective ventilation is the main defense in the low cost house using OPS concrete by reducing ceiling temperatures. OPS aggregate being light weight and porous, and OPS hollow block having 32% void act as thermal insulation with lower thermal conductivity compared with conventional concrete [5]. However, in a low cost house, using OPS concrete, the roof design and mainly ventilation seems to be a very important parameter in reducing the attic temperature and the most cost effective tool in passive cooling for tropical environments. Figures 8.45 shows the latest condition of the low cost house built using OPS concrete.

Table 8.6. Measured temperatures for three different houses during 28 days from 9 July 2004 to 5 August 2004 [6]

Sl	House description	Atmospheric temp (^{0}C)		Room temp. (^{0}C)		Attic temp. (^{0}C)	
		Min.	Max.	Min.	Max.	Min.	Max.
1	House-1: Conventional one story house with built up area of 69.39m^2,	22.5	37.9	26.4	33.6	23.2	44.9
		Difference = 15.4		Difference = 7.2		Difference = 21.7	
2	House-2: Conventional one story house with built up area of 79.17m^2	22.48	37.88	25.56	37.44	21.33	54.13
		Difference = 15.4		Difference = 11.88		Difference = 32.8	
3	Low cost house built using OPS lightweight concrete with built up area of 58.68m^2	22.5	36.6	26.3	31.1	22.9	35.3
		Difference = 14.1		Difference = 4.8		Difference = 12.4	

Figure 8.45. Low cost house built using OPS concrete, latest photo [courtesy, Dr. Mohd Kamal Mohd Shah, personal contact].

8.4. More Than 5 Years in Sustained Load onto a Precast C-Channel

The full precast RC floor panel known as a C-channel (Figure 8.46) with 1m width and 3.1m long was designed using an imposed load of 1.5kN/m^2 and cast with OPS concrete as the residential building floor. The 28-day compressive strength of companion cubes were recorded at 27 N/mm^2. It was cast outside the laboratory to suit the site condition. It was erected onto supports using a forklift at the age of 28-days to verify the carrying capacity of an imposed load under a long sustained period of time (Figure 8.47). Each layer was loaded with 134 clay bricks. The average weight of each clay brick at dry condition was 2.2 kg. The dial gauges, two near the supports and the rest at centre span were placed underneath the panel to measure the deflections. Thus, 12 layers of bricks were placed up to the deflection limit (span/250). At mid span, a 50mm gap was provided to avoid arch action by the bricks. Any additional load occurring due to water absorption in the bricks during rain was not considered. The physical monitoring was done for more than 5 years to notice any defects or crack formations either on ribs or flanges (Figures 8.48 and 8.49).

Figure 8.46. Precast C-channel made using OPS concrete with 3.1 m long and 1.0 m width lifted using a forklift [7].

A comparative study using conventional normal weight concrete was also conducted using precast C-channels (Figures 8.50 to 8.52). The long term performance was also physically monitored using same size C-channel made with normal weight concrete (Figures 8.50 to 8.52). This was also cast, erected and loaded outside the laboratory similar to C-channels made with OPS concrete (Figures 8.46 to 8.49). A total of 134 dry clay bricks were placed at each layer. It was loaded with 16 layers of dry clay bricks (Figures 8.50 and 8.51) up to deflection limit (span/250). The dial gauges were also placed to measure the deflection during loading with bricks. The tests using UDL were conducted outside the laboratory, so rainwater or moisture changes in the bricks were not taken into consideration. The companion concrete cubes were tested at 28-days for compressive strength of 30 N/mm^2. It remained for more than 5 years for physical monitoring of any defects or cracks that may have occurred either on the ribs or flange soffit. Based on more than 5 years of sustained loading, both precast C-channels made with OPS concrete and normal weight concrete showed no sign of defect or crack formation.

Figure 8.47. 3.0 m span and 1.0 m width precast floor panel, C-channel made with OPS concrete of 27 N/mm^2 and loaded with 1,608 clay bricks of 35.38kN (11.79 kN/m^2 floor area) at deflection limit in February 2007. A 50mm gap was provided to avoid arch action of the bricks. The panel made with two ribs, each with 160mm deep and flange with 75mm thick [7].

Figure 8.48. No sign of defects and crack formation noticed through visual observation over more than 5 years with sustained load of this precast floor panel. The C-channel was made with OPS concrete [courtesy, author].

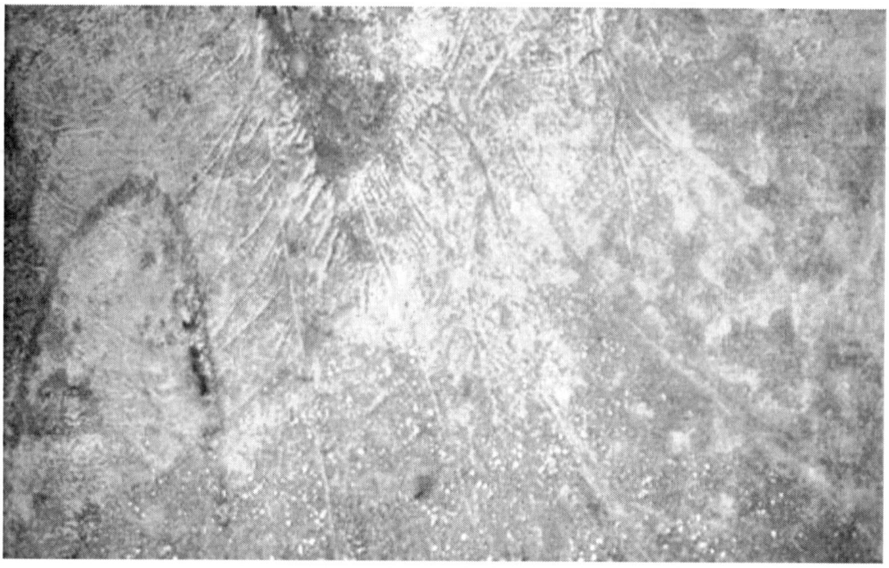

Figure 8.49. No sign of defects seen on the slab soffit of precast floor panel. C-channel was made using OPS concrete having sustained load shown above. This photo taken on 14 April 2012 [courtesy, author].

Figure 8.50. 3.0 m span and 1.0 m width precast floor panel known as a C-channel, made with normal weight concrete of 30 N/mm^2 comp. st. at 28-days and loaded with 2,144 clay bricks of 47.17kN (15.72kN/m^2) at deflection limit in December 2006. A 50mm gap was provided to avoid arch action of bricks. Panel made with two ribs each with 160mm deep and flange with 75mm thick [7].

Figure 8.51. No sign of defects and crack formations were noticed through visual observation over more than 5 years with sustained load of this precast floor panel. The C-channel made with normal weight concrete. This photo was taken on 14 April 2012 [courtesy, author].

Figure 8.52. No sign of defects seen on the slab soffit of precast floor panel. The C-channel was made using normal weight concrete having sustained load shown above. This photo taken on 14 April 2012 [courtesy, author].

8.5. Chapter Remarks

It is concluded that OPS concrete as structural lightweight concrete shows a wonderful structural performance in a footbridge over 15 years. It performs well for low cost houses, in both structural and thermal comfort aspects. It has not shown any structural defects during a sustained load period of more than 5 years.

References

[1] Research Report. 2001. *Reinforced concrete slab made of OPS (Oil Palm Shell) concrete, Grant no. PP010/2000.* Universiti Malaysia Sabah.

[2] *Oil Palm in Malaysia-E-tawau*, www.etawau.com/OilPalm/OilPalm.htm. Retrieved on 25 October 2016.

[3] Research Reports. 2004. *Low cost house using OPS lightweight concrete, Grant no. B-03-02-02-PR/007 and Housing for low-income families in East Malaysia, Grant no. British council, Malaysia. no. HEL 1205.* Universiti Malaysia Sabah.

[4] MS1525: 2007, *Malaysian Standard*, Code of practice on energy efficiency and use of renewable energy for non-residential building, first revision.

[5] R.W.Steiger, M.K. Hurd, 1978, Lightweight insulating concrete for floors and roof decks, *Concrete Construction.* **23**(7). pp 411-422.

[6] H. DJamila 2006. *www.eco-web.com/edi/060927.html.* Retrieved on 20 April 2016.

[7] Research Report. 2009. *Semi-precast lightweight flooring slab using OPS, CIDB Project no. LPIPM: CREAM/UPP04-02-10-04-11.* Part-3. Universiti Malaysia Sabah.

Chapter 9

PRECAST SIConSofa BLOCK FOR MULTIFUNCTIONS

Keywords: Precast SIConSofa concrete block, Precast RC Bedding block, Water Seepage drain, Decorative Retaining wall, Slope surface protection

9.1. INTRODUCTION

The non-structural precast concrete block, SIConSofa like the shape of a 'home-sitting-sofa' being self-interlocked is used in assembling the retaining wall to retain the soil. This SIConSofa is an award winning product at the 33rd International Exhibition of Inventions, New Techniques and Products in Geneva, Switzerland, 6 10 April 2005. It is made using either OPS concrete or conventional normal weight concrete. A total of 55 SIConSofa blocks are made using $1m^3$ concrete which covers a wall surface of $5.6m^2$. In another way, 9.8 SIConSofa blocks can cover a $1m^2$ area. It is estimated that a 30-$35m^2$ wall area of SIConSofa blocks can be assembled daily by two skilled workers if a simple lifting device is used. Using full precast SIConSofa and bedding RC blocks, the man-days involved for installation is about one-third of that of the conventional practice in RC retaining walls. The exposed surfaces of SIConSofa blocks in retaining walls and slope surface protection show a more aesthetic appearance. The drain assembled with precast RC bedding blocks and these SIConSofa blocks in walls enhances water discharge through side and bottom seepage during rain as the ground water table is quite low, especially in summer season. This is a potential product to eliminate mosquito breeding in drains. Thus, this system is the optimum choice for applications such as (i) water seepage drains (ii) decorative retaining walls, and (iii) soil slope surface coverings.

This chapter discusses the shape of SIConSofa. It explains precast bedding blocks for different widths of drains. How road side drains, particularly water seepage drains and main drains can be built are also discussed. It discusses how these blocks fulfil the requirement of a retaining wall structure up to 3m high with different surcharge loads. The six years in service of drains, cum decorative retaining built using SIConSofa and precast RC bedding blocks are elaborated. Finally this chapter discusses conceptually how soil slope surfaces can be covered with these SIConSofa blocks.

9.2. Shape of Precast SIConSofa Concrete Blocks

The shape of a SIConSofa concrete block has been determined based on single unit sitting sofa used in a home or office. The grooves are provided for interlocking systems. The shape is shown in Figures 9.1 to 9.3. This block is manually lifted as shown in Figure 9.4. It is suggested to use a simple lifting device in assembling at construction site.

Figure 9.1. Prototype SIConSofa concrete block [1].

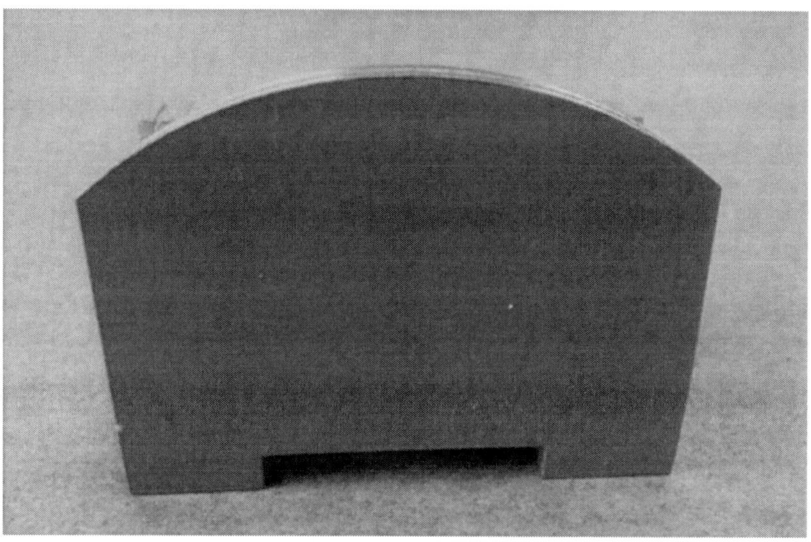

Figure 9.2. Exposed surface of a SIConSofa model block [1].

(a) Ridge of SIConSofa model block

(b) Ridgc of Prototype SIConSofa block

Figure 9.3. Central ridge for an interlocking system of SIConSofa model blocks [1].

Figure 9.4. Manually lifted SIConSofa block [1].

9.3. Precast RC Bedding Blocks for Different Widths of Drains

The drain system using SIConSofa blocks is made in a trapezoidal shape, where the bottom width is smaller than the top width. The drain widths at the bottom from 300mm to 700mm are proposed using the same size of precast RC bedding blocks with a 75mm bearing width as shown in Figures 9.5 to 9.8. The bearing width of more than 1,000mm bottom width of such a drain is proposed with 100mm due to water load (Figures 9.9 to 9.12). The RC bedding blocks are proposed of 475mm width (Figures 9.9 to 9.12). Figure 9.13 shows the perspective view on how this drain system is assembled including bedding blocks. The mock drain system up to 1m deep is assembled on flat ground to observe the interlocking system (Figures 9.14 and 9.16).

There are three types of precast components: (i) RC bedding blocks, (ii) Flat RC blocks sitting on bedding blocks and (iii) SIConSofa. The length of a precast bedding block is fixed at 600mm long. Compacted granular materials between two bedding blocks are required to adjust as per requirements. The ends of each bedding block has two grooves for interlocking with adjacent bedding blocks. The interlocking system between two adjacent bedding blocks at site can be done using either precast concrete blocks or cast in-situ concrete.

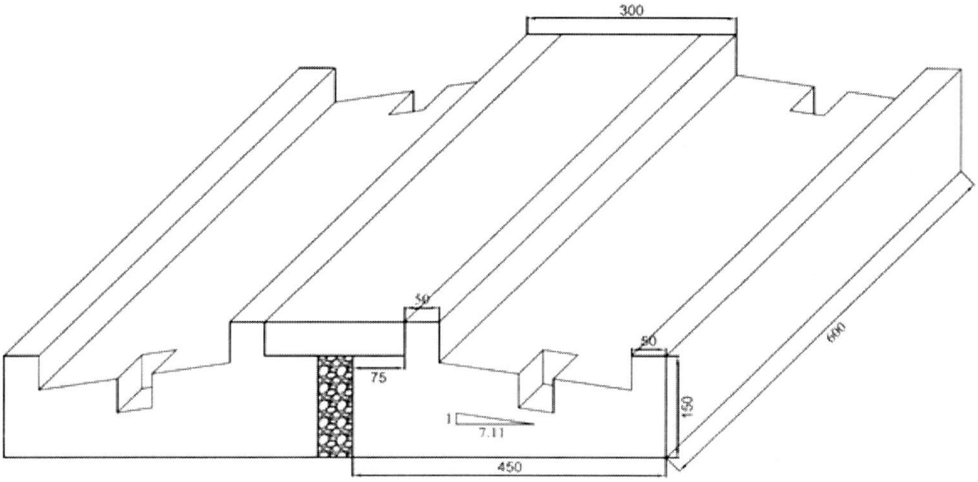

Figure 9.5. Perspective view of precast RC bedding blocks for 300mm width drain. Each precast bedding block is 600mm long with two grooves for interlocking system at both ends [courtesy, author].

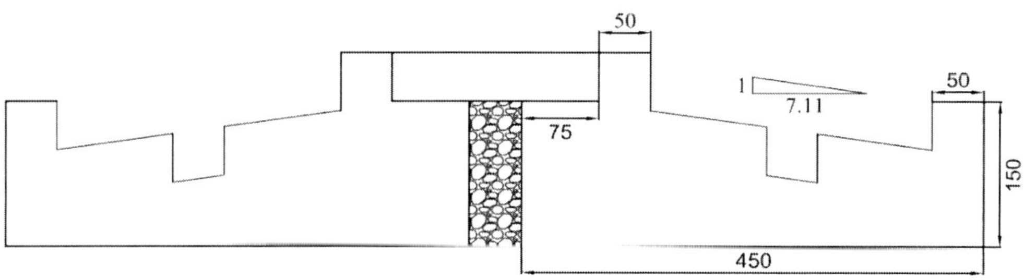

Figure 9.6. Cross section of precast RC bedding blocks for 300mm width drain. Each precast bedding block has a 450mm width including 75 mm bearing and 600mm long with two grooves for interlocking system at both ends [courtesy, author].

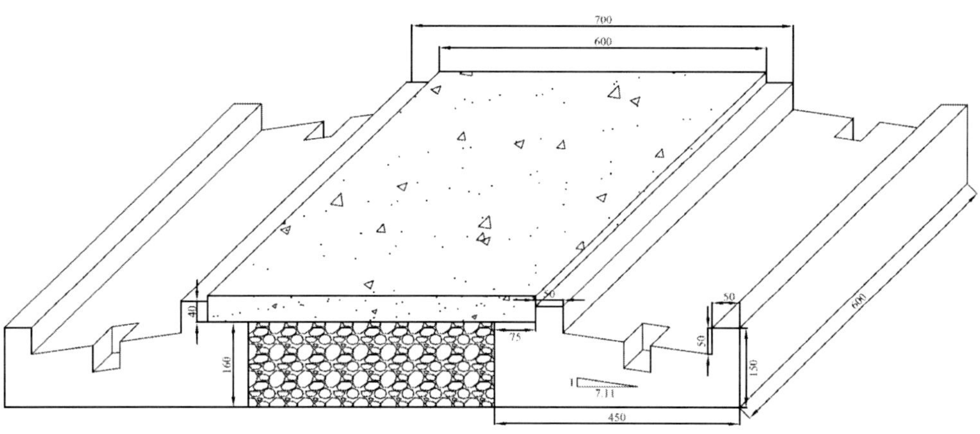

Figure 9.7. Perspective view of precast RC bedding blocks for 700mm width drain. Each precast bedding block is 600mm long with two grooves for interlocking system at both ends [courtesy, author].

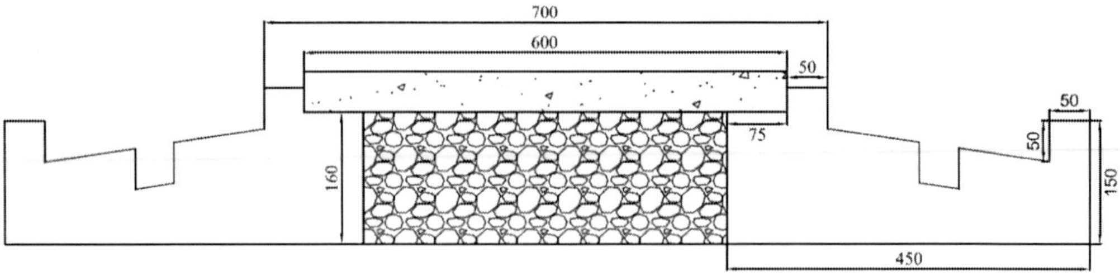

Figure 9.8. Cross section of precast RC bedding blocks for 700mm width drain. Each precast bedding block has a 450mm width including 75 mm bearing and 600mm long with two grooves for interlocking system at both ends [courtesy, author].

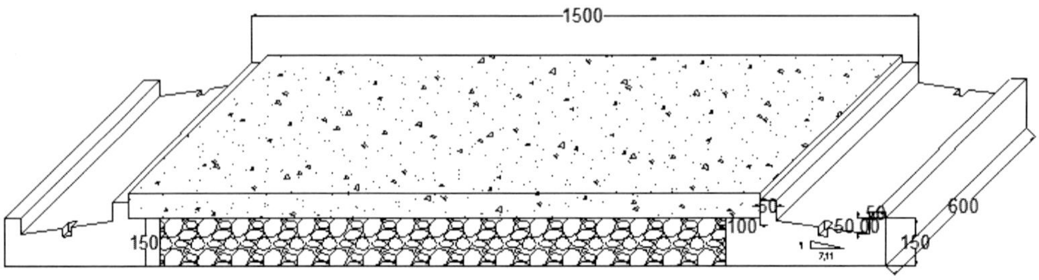

Figure 9.9. Perspective view of precast RC bedding blocks for 1,500mm width drain. Each precast bedding block is 600mm long with two grooves for interlocking system at both ends [courtesy, author].

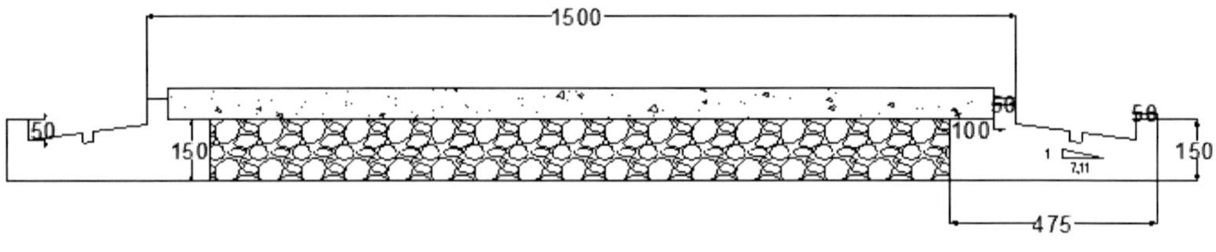

Figure 9.10. Cross section of precast RC bedding blocks for 1,500 mm width drain. Each precast bedding block has a 475 mm width including 100 mm bearing and 600mm long with two grooves for interlocking system at both ends [courtesy, author].

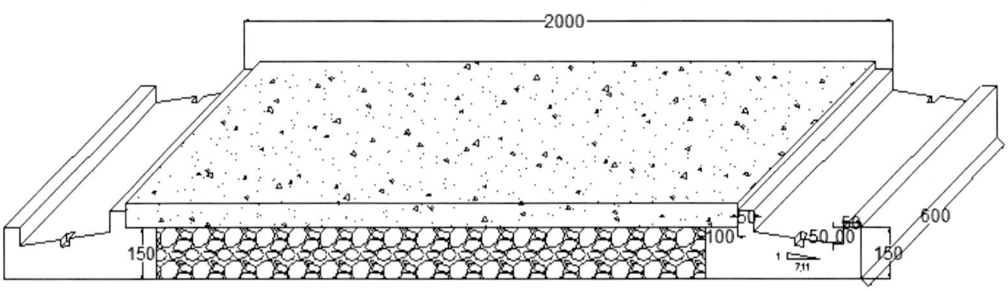

Figure 9.11. Perspective view of precast RC bedding blocks for 2,000 mm width drain. Each precast bedding block is 600mm long with two grooves for interlocking system at both ends [courtesy, author].

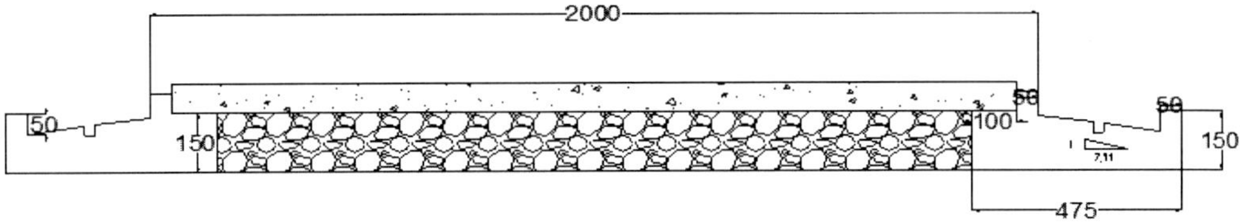

Figure 9.12. Cross section of precast RC bedding blocks for 2,000 mm width drain. Each precast bedding block has a 475 mm width including 100 mm bearing and 600mm long with two grooves for interlocking system at both ends [courtesy, author].

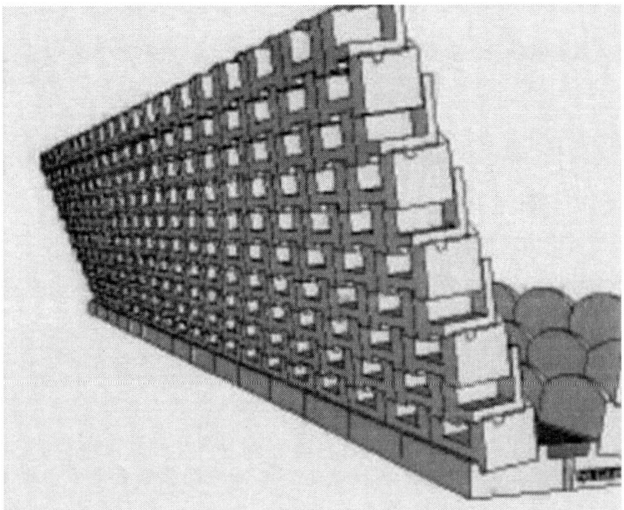

Figure 9.13. Perspective view of typical arrangement of SIConSofa forming drain system with precast RC bedding blocks [courtesy, author].

Figure 9.14. Mock drain of about 1m deep assembled using SIConSofa blocks made with OPS concrete and cast in-situ concrete as flat bedding [1].

Figure 9.15. Mock drain with typical arrangement with SIConSofa forming 7m long and about 1m deep with cast in-situ concrete as flat bedding, half precast SIConSofa assembled also seen at end [1].

Figure 9.16. Mock drain system with drain cover using SIConSofa, drain cover loaded with precast wall blocks and concrete slab of $2kN/m^2$ [1].

9.4. WATER SEEPAGE DRAIN IN HOUSING AREA USING SICONSOFA BLOCKS

Mosquitoes cause serious problems to humans and other animals globally. They breed in standing shallow water that stands more than three days. The open drains in flat land cause stagnant water favoring the breeding of mosquitoes and other insects. The best way to eliminate mosquitoes and other insects is to eliminate the breeding sites. Figures 9.17 to 9.19 show the conventional drain systems which are filthy and favorable for the breeding of mosquitoes and other insects. The occurrence of filthy conditions in such drains is due to the discharge of waste water from kitchens and showers. Especially waste water from kitchens is mixed with waste food that creates filthy and favorable conditions for germs to grow in conventional drains. The author has noticed strong odors in such drains. People also throw rubbish into such drains which also prevents flow (Figures 9.17 and 9.18). This conventional drain is made normally with a very gentle gradient in flat ground which does not allow to flow properly especially during non-raining period.

The new drain system to be known as the 'Water Seepage Drain' using interlocking concrete blocks is proposed to avoid the problem seen in conventional drains as discussed above. The SIConSofa, an IBS product is used as a retaining wall without using any mortar to retain earth. The SIConSofa blocks satisfy the requirements of retaining wall structures. The drain assembled with interlocking precast RC bedding blocks and SIConSofa blocks enhances water discharge through side and bottom seepage especially when the ground water table is low. This new drain does not allow the build up of any pore pressure at the back of the wall. Figures 9.20-9.22 show the proposed Water Seepage Drain assembled with precast SIConSofa and RC bedding blocks. This new drain does not hold water, so it is able to eliminate the breeding of mosquitoes and other insects, and it also claims to minimize the occurrence of odors.

Figure 9.23 shows the conventional main drain built with rubbles as cast in-situ. This type of main drain also can be assembled using SIConSofa blocks and RC bedding blocks as proposed in Figure 9.24.

Figure 9.17. Filthy drain favoring the breeding of mosquitos and other insects seen very commonly in conventional drains [courtesy, author].

Figure 9.18. Roadside drain seen here with countless insect-larvae in stagnant water and strong odor, a serious health risk environment in housing areas [courtesy, author].

Figure 9.19. Conventional drain with stagnant water - a very common problem due to impervious drain surface [courtesy, author].

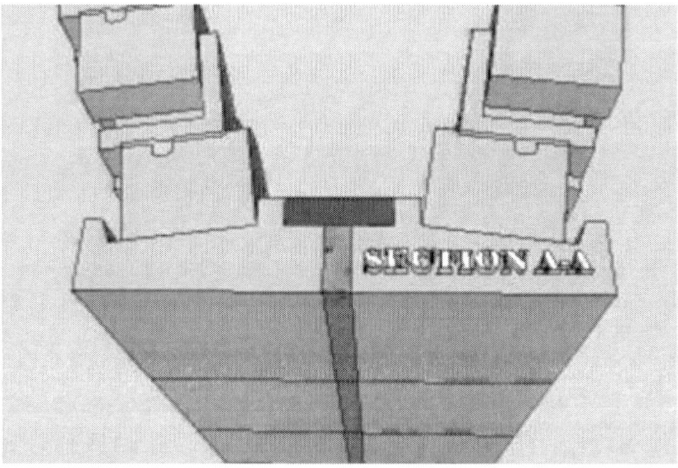

Figure 9.20. Bottom view of proposed 'Water Seepage Drain' assembled using precast SIConSofa and RC bedding blocks allowing water seepage through block joints downward and laterally [courtesy, author].

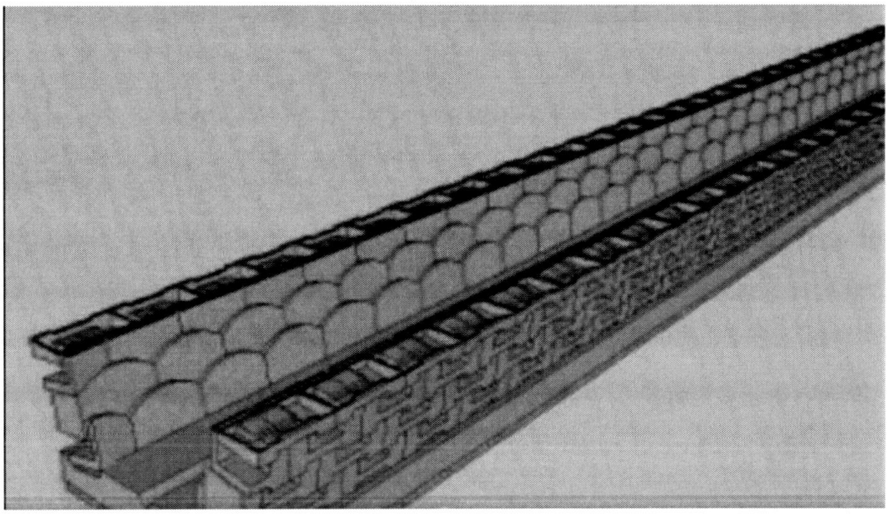

Figure 9.21. Perspective view of proposed 'Water Seepage Drain' assembled using precast SIConSofa and RC bedding blocks while keeping top level flat [courtesy, author].

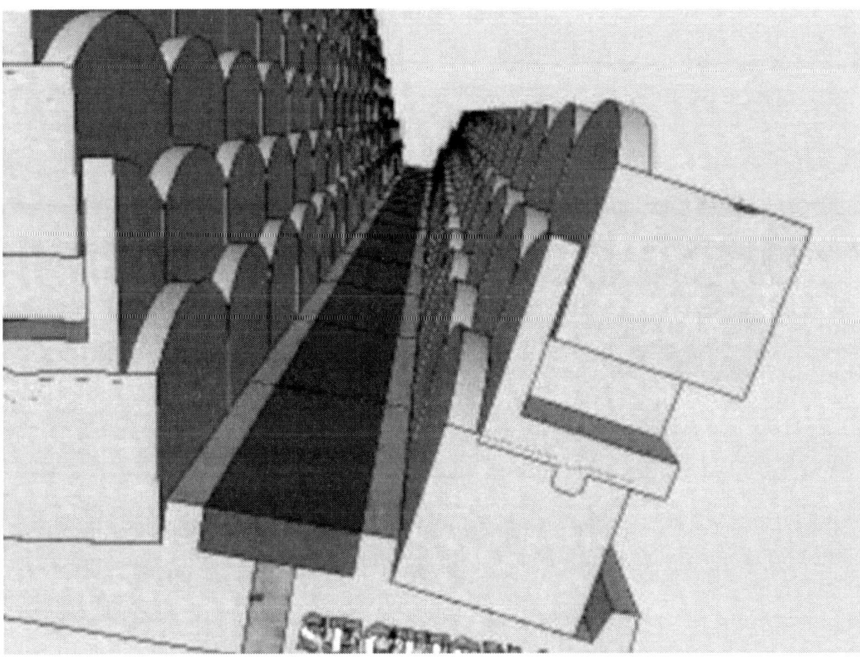

Figure 9.22. Internal view of proposed 'Water Seepage Drain' assembled using precast SIConSofa, RC bedding blocks, flat RC panel sitting on RC bedding blocks and granular filling [courtesy, author].

Figure 9.23. Conventional main drain of 2.75m deep and 6.15m wide at top made using rubbles [1].

Figure 9.24. Use of precast SIConSofa and RC bedding blocks for main drain with adjustable bottom and top widths for required discharge - an alternative approach to conventional main drain made with rubbles as shown in Figure 9.23 [courtesy, author].

9.5. Decorative Retaining Wall Using SIConSofa Blocks

9.5.1. Forces Acting on Retaining Wall

The forces that act on the retaining wall are:

1) Those forces that tend to cause the wall to move and
2) Those forces that oppose movement of the wall.

Figure 9.25 illustrates the various forces acting on the retaining wall built using SIConSofa blocks.

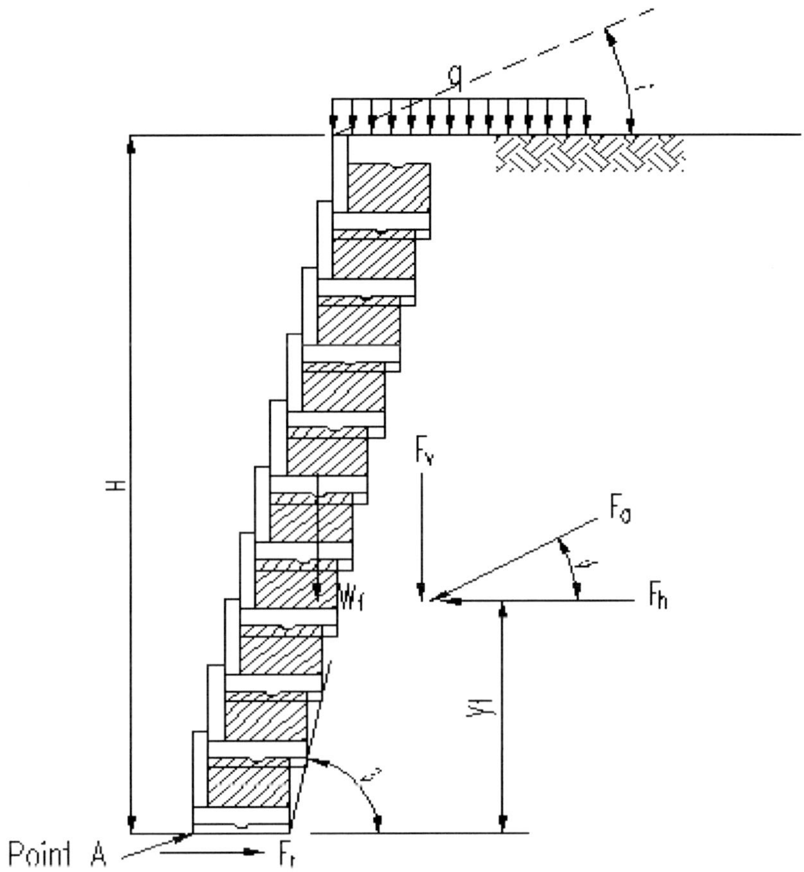

Figure 9.25. Illustration of various forces acting on retaining wall using SIConSofa blocks [1].

Where;

F_a = Active force on the retaining wall; it is the resultant force of the active pressure on the retaining wall.

q = Surcharge load

F_v = Vertical component of the active force

F_h = Horizontal component of the active force

F_r = Maximum frictional resistance to sliding

W_f = Weight of the wall facing

H = Distance from bottom of the wall to the top of the wall

γ = Unit weight of the soil

ϕ = Friction angle of the soil

β = Angle between the horizontal and the sloped back face of the wall

i = Slope of the top of the retained soil

ϕ_w = Angle between a line perpendicular to the wall face and the line of action of the active force.

C_f = Coefficient of friction

9.5.2. Pressure Coefficient

The horizontal stress (σ_h) on the retaining wall due to the retained soil is directly proportional to the vertical stress (σ_v) on the soil at the same depth. The ratio of the two stresses is a constant called the pressure coefficient:

$$K = \frac{(\sigma_h)}{(\sigma_v)}$$

The active pressure coefficient can be calculated using Coulomb's equation. This equation takes into account the slope of the backfill, the batter of the retaining wall, and the effects of friction between the retained soil and the surface of the retaining wall.

Where;

ϕ = the friction angle of the soil

β = angle between the horizontal and the sloped back face of the wall

i = slope of the top of the retained soil

ϕ_w = angle between a line perpendicular to the wall face and the line of action of the active force

$$= \frac{2}{3}\phi$$

$$K_a = \left[\frac{\csc(\beta)\sin(\beta-\phi)}{\sqrt{\sin(\beta+\phi_w)} + \sqrt{\frac{\sin(\phi+\phi_w)\sin(\phi-i)}{\sin(\beta-i)}}}\right]^2 \quad [2]$$

9.5.3. Active Force on the Wall

The active pressure distribution is triangular, which reflects the fact that soil pressure increases linearly with soil depth. The vertical pressure at any depth is given by:

$$P_v = (\gamma)(H)$$

Where;

γ = unit weight of the soil

H = distance from the bottom of the wall to the top of the wall

The horizontal pressure (P_h) is related to the vertical pressure (P_v) by the active pressure coefficient.

$$K_a = \frac{(P_h)}{(P_v)}$$

Since K_a and γ are constants, the horizontal pressure increases linearly as the depth increases and the resulting pressure distribution is triangular. The magnitude of the resultant force of a triangular pressure distribution is equal to the area of the triangle. The pressure at the base of the triangle is given by:

$$P_h = (K_a)(\gamma)(H)$$

The magnitude of the active force is:

$$\begin{aligned} F_a &= \text{area of the triangle} \\ &= (0.5)(\text{base})(\text{height}) \\ &= (0.5)(P_h))(H) \\ &= (0.5)(K_a)(\gamma)(H)(H) \\ &= (0.5)(K_a)(\gamma)(H)^2 \end{aligned}$$

Therefore, the equation for the active force on the retaining wall is:

$$F_a = (0.5)(K_a)(\gamma)(H)^2$$

The resultant force acts at a point above the base equal to $\frac{1}{3}$ of the height of the triangle. Adding a surcharge or slope above the wall has the effect of adding a rectangular pressure distribution. The resultant force of a rectangular pressure distribution acts at a point above the base equal to $\frac{1}{2}$ of the height of the rectangle.

9.5.4. Calculation of Effective Unit Weight of Wall Facing

The effective unit weight of the wall facing is often needed for wall design. A modular concrete block's unit weight is the sum of the blocks plus the wall refill material and is calculated below. Concrete usually weighs more than soil. A typical unit weight for non-reinforced mass concrete is 22.6 kN/m^3 while a typical unit weight for soil (gravel) is 19.65 kN/m^3.

The weight of one unit SIConSofa concrete block is approximately 40 kg (0.3924 kN). The unit weight of the concrete is 22.6 kN/m^3. The block dimensions used are: Length (l) = 0.45 m, Height (h) = 0.305 m and Depth (t) = 0.25 m. From these values, the volume of concrete for each SIConSofa block can be calculated:

$$V_c = \frac{0.3924\ kN}{22.6\ kN/m^3} = 0.017\ m^3$$

The total volume occupied by each standard SIConSofa block unit, including the voids, is:

$V_{tot} = (0.45\ m)\ (0.305\ m)\ (0.25\ m)$
$= 0.034\ m^3$

Therefore, the volume of voids is:

$V_v = 0.034\ m^3 - 0.017\ m^3$
$= 0.017\ m^3$

The unit weight of the wall facing can now be calculated. Assuming that the voids will be filled with wall sand with a unit weight of 19.65 kN/m^3, the unit weight of the wall facing is:

$$\gamma_{wall} = \frac{(weight\ \ of\ \ concrete) + (weight\ \ of\ \ wall\ \ soil)}{(volume\ \ of\ \ block)}$$

$$= \frac{(weight\ \ of\ \ concrete) + (weight\ \ of\ \ wall\ \ soil)}{(V_{tot})}$$

$$= \frac{(0.017\ \ m^3)(22.6\ \ kN/m^3) + (0.017\ \ m^3)(19.65\ \ kN/m^3)}{(0.034\ \ m^3)}$$

$= 21.13\ kN/m^3$

Once the unit weight of the wall facing is known, it is a simple matter to calculate the weight per linear meter of wall:

W_f = (unit weight of wall) (volume of wall)
= (unit weight of wall) (wall height) (facing depth)
$= (\gamma_{wall})\ (V_w)$
$= (\gamma_{wall})\ (H)\ (t)$

For a wall 1.767 m high with a facing depth of 0.25 m, the weight of the facing would be:

$W_f = (21.13\ kN/m^3)\ (1.767\ m)\ (0.25\ m)$
$= 9.33\ kN/m$

9.5.5. Safety Factors

The safety factors used in this design manual conform to the guidelines of the Federal Highway Administration, Mechanically Stabilized Earth Walls and Reinforced Soil Slopes – Design and Construction Guidelines. They recommend using the following safety factors;

Sliding > 1.5
Overturning > 1.5

9.5.6. Simple Gravity Wall

A simple gravity wall relies on the weight of the wall to counteract the force of the retained soil. Two modes of failure must be analyzed, 1) sliding and 2) overturning.

9.5.7. Sliding Failure

A simple gravity wall will not fall by sliding if the force resisting. F_r, is greater than or equal to the force causing sliding, F_h. The force resisting sliding is the frictional resistance at the wall base. The minimum safety factor for sliding failure is 1.5. Therefore, F_r, must be greater than or equal to (1.5) F_h.

For a 10 course wall, given:

$\phi = 40°$

$i = 0°$

$\beta = 70°$

$$K_a = \left[\frac{\csc(\beta)\sin(\beta-\phi)}{\sqrt{\sin(\beta+\phi_W)+\sqrt{\dfrac{\sin(\phi+\phi_W)\sin(\phi-i)}{\sin(\beta-i)}}}}\right]^2$$

$$= \left[\frac{\csc(70°)\sin(70°-40°)}{\sqrt{\sin(70°+26.67°)+\sqrt{\dfrac{\sin(40°+26.67°)\sin(40°-0°)}{\sin(70°-0°)}}}}\right]^2$$

$$= \left[\frac{(1.0642)(0.5)}{0.9966+0.7925}\right]^2$$

$$= 0.0885$$

$H = 2.06$ m

$\gamma = 19.65\ kN/m^3$ (assumed pressure due to materials)

$$\phi_w = \frac{2}{3}\phi$$

$$= 26.67°$$

$$y_t = \frac{1}{3}H$$

$$C_f = \tan(\phi)$$

$$= 0.8391$$

The total active force exerted by the soil on the wall:

$$F_a = (0.5)(K_a)(\gamma)(H)^2$$

$$= (0.5)(0.0885)(19.65\ kN/m^3)(1.767\ m)^2$$

$$= 2.71\ kN/m$$

The active force can be resolved into a component perpendicular to the wall and a component parallel to the wall. The degree of the angle between the active force and a line perpendicular to the face of the wall is ϕ_w. ϕ_w varies according to the compaction level of the soil. For very loose soil ϕ_w approaches ϕ; for compacted soil, ϕ_w can be as low as $\frac{2}{3}\phi$. Since the wall design involves compacting the backfill soil, the value of $\phi_w = \frac{2}{3}\phi$. Thus, the horizontal component of the active force is:

$$\begin{aligned} F_h &= (F_a)\cos(\phi_w) \\ &= (2.71\ \text{kN/m})\cos(26.67°) \\ &= 2.42\ \text{kN/m} \end{aligned}$$

Similarly, the vertical component of the active force is:

$$\begin{aligned} F_v &= (F_a)\sin(\phi_w) \\ &= (2.71\ \text{kN/m})\sin(26.67°) \\ &= 1.22\ \text{kN/m} \end{aligned}$$

The weight of the wall facing must be determined before the frictional resistance to sliding can be calculated:

$$\begin{aligned} W_f &= (21.13\ \text{kN/m}^3)(1.767\ \text{m})(0.25\ \text{m}) \\ &= 9.33\ \text{kN/m} \end{aligned}$$

The maximum frictional resistance to sliding, F_r is calculated by multiplying the total vertical force, V_t, by the coefficient of friction. The total vertical force is the sum of the weight of the wall and the vertical component of the active force. The coefficient of friction, C_f, is assumed to be equal to tan (ϕ). Thus, the maximum frictional resistance is:

$$\begin{aligned} F_r &= (V_t)(C_f) \\ &= (V_t)\tan(\phi) \\ &= (W_f + F_v)\tan(\phi) \\ &= (9.33\ \text{kN/m} + 1.22\ \text{kN/m})\tan(40°) \\ &= 8.85\ \text{kN/m} \end{aligned}$$

Finally, the safety factor against sliding can be calculated:

$$\begin{aligned} SFS &= \frac{(Force\ \ resisting\ \ sliding)}{(Force\ \ cau\sin g\ \ sliding)} \\ &= \frac{(F_r)}{(F_h)} \end{aligned}$$

$$= \frac{(8.85 \ kN/m^3)}{(2.42 \ kN/m^3)}$$

$= 3.7 > 1.5$ OK

The safety factor against sliding is more than 1.5. Therefore, the wall is stable and doesn't require reinforcement to prevent sliding failure. However, the wall against overturning failure must be analyzed.

9.5.8. Overturning Failure

Overturning failure occurs when the forces acting on the wall cause it to rotate about the bottom front cover of the wall (Point A). For stability, the moments resisting overturning, M_r, must be greater than or equal to the moments causing overturning, M_o. The minimum safety factor for overturning is 1.5. Therefore, M_r must be greater than or equal to (1.5) M_o.

$$Mr=(W_f)\ [(t/2) + (0.5)\ (H)\ \tan\ (90° - \beta)] + (F_v)\ [(t) + \frac{1}{3}(H)\ \tan\ (90° - \beta)]$$

$$=(9.33\ kN/m)[(0.125\ m) + (0.5)\ (1.767\ m)\ \tan\ (90° - 70°)] + (1.22\ kN/m)[(0.25\ m) + \frac{1}{3}(1.767\ m)\ \tan\ (90° - 70°)]$$

=4.17 kN-m/m + 0.57 kN-m/m

=4.74 kN-m/m

The horizontal component of the active force is the only force that contributes to the overturning moment. The active force is the result of the active pressure distribution, which is triangular. For triangular pressure distributions, the vertical centroid is located at $\frac{1}{3}$ the height of the triangle. Therefore, the horizontal component of the active force acts on the wall $\frac{1}{3}$(H) from the bottom of the wall, where $y_t = \frac{1}{3}H$. The moment causing overturning is:

$M_o = (F_h)\ (y_t)$

$= (2.42\ kN/m)\ (\frac{1}{3})\ (1.767\ m)$

$= 1.43\ kN/m$

The safety factor against overturning is:

$$SFO = \frac{(Moment \quad resisting \quad overturning)}{(Moment \quad causing \quad overturning)}$$

$$= \frac{(M_r)}{(M_o)}$$

$$= \frac{(4.74 \quad kN-m/m)}{(1.43 \quad kN-m/m)}$$

$= 3.3 > 1.5$ OK

The safety factor against overturning is more than 1.5. Therefore, the wall is stable and it does not need geogrid reinforcement to prevent overturning. The safety factor against sliding is also greater than 1.5 for this wall. This wall is adequately reinforced againt sliding and overturning and no geogrid reinforcement is required.

9.5.9. Surcharge Load on Simple Gravity Walls

The surcharge load of 10 kN/m^2 is estimated for the analysis. The surcharge is located next to the back of the wall. The calculation of the pressure on the retaining wall due to surcharge is:

$P_q = (q)\,(K_a)$
$= (10\ kN/m^2)\,(0.0885)$
$= 0.89\ kN/m^2$

Therefore, the calculation for the horizontal and vertical component of the pressure is:

$P_{qh} = (P_q)\cos(\phi_w)$
$= (0.89\ kN/m^2)\cos(26.67°)$
$= 0.8\ kN/m^2$
$P_{qv} = (P_q)\sin(\phi_w)$
$= (0.89\ kN/m^2)\sin(26.67°)$
$= 0.4\ kN/m^2$

The total surcharge forces on the wall are calculated as:

$F_{qh} = (P_{qh})\,(H)$
$= (0.8\ kN/m^2)\,(1.767\ m)$
$= 1.41\ kN/m$

$F_{qv} = (P_{qv})\,(H)$
$= (0.4\ kN/m^2)\,(1.767\ m)$

$= 0.71$ kN/m

For a simple gravity wall, the horizontal force due to the surcharge is a force that tends to cause both sliding and overturning. Therefore, it must be added to those forces when the safety factors are calculated.

The safety factor against sliding is:

$$\text{SFS} = \frac{F_r + (F_{qv})(C_f)}{F_h + F_{qh}}$$

$$= \frac{(8.85\ kN/m^2) + (0.71\ kN/m^2)\tan(40°)}{(2.42\ kN/m^2) + (1.41\ kN/m^2)}$$

$= 2.5 \quad > \quad 1.5$ OK

The safety factor against overturning is:

$\Sigma M_r = (W_f)\ [(t/2) + (0.5)\ (H)\ \tan\ (90° - \beta)] + (F_v)\ [(t) + \frac{1}{3}(H)\ \tan\ (90° - \beta)] + (F_{qv})\ [(t) + (0.5)\ (H)\ \tan\ (90° - \beta)]$

$= (9.33\ \text{kN/m})[(0.125\ \text{m}) + (0.5)\ (1.767\ \text{m})\ \tan\ (90° - 70°)] + (1.22\ \text{kN/m})[(0.25\ \text{m}) + \frac{1}{3}(1.767\ \text{m})\ \tan\ (90° - 70°)] + (0.71\ \text{kN/m})[(0.25\ \text{m}) + (0.5)\ (1.767\ \text{m})\ \tan\ (90° - 70°)]$

$= 4.17$ kN-m/m + 0.57 kN-m/m + 0.41 kN-m/m

$= 5.15$ kN-m/m

$\Sigma M_o = (F_h)\ (y_t) + (F_{qh})\ (0.5)\ (H)$

$= (2.42\ \text{kN/m})\ (\frac{1}{3})\ (1.767\ \text{m}) + (1.41\ \text{kN/m})\ (0.5)\ (1.767\ \text{m})$

$= 1.43$ kN/m + 1.25 kN/m

$= 2.68$ kN/m

$$\text{SFO} = \frac{(\Sigma M_r)}{(\Sigma M_o)}$$

$$= \frac{(5.15\ kN-m/m)}{(2.68\ kN-m/m)}$$

$= 1.9 > 1.5$ OK

The surcharge on the backfill will lower the safety factors compared to the recommended minimum values of 1.5 for both sliding and overturning. The surcharge can make the difference between a stable wall and an unstable wall. Table 9.1 shows different wall heights, slope angles, surcharge loads with safety factors against sliding and overturning. Figure 9.26

shows the proposed retaining wall using SIConSofa blocks for repair work of a damaged road shoulder.

Table 9.1. Safety factors of decorative retaining walls considering different wall heights, slope angles and surcharge loads [courtesy, author]

H (m)	0.5		1.0		1.5		2.0		2.5		3.0			3.0					
ß (degree)	70													68			65		
Q (kN/m^2)	0	10	0	10	0	10	0	10	0	10	0	5	10	0	5	10	0	5	10
SFS >1.5	12	4.4	6.1	3.3	4.2	2.7	3.3	2.3	2.7	2.0	2.3	2.0	1.8	2.5	2.2	2.0	2.9	2.6	2.3
SFO >1.5	18.6	5.0	6.8	3.0	4.0	2.2	2.9	1.8	2.2	1.5	1.8	1.5	1.3	2.2	1.8	1.5	2.8	2.3	1.9

H- wall height
β-angle between horizontal and sloped back face of wall
Q- surcharge load
SFS-Safety factor against sliding
SFO-Safety factor against overturning

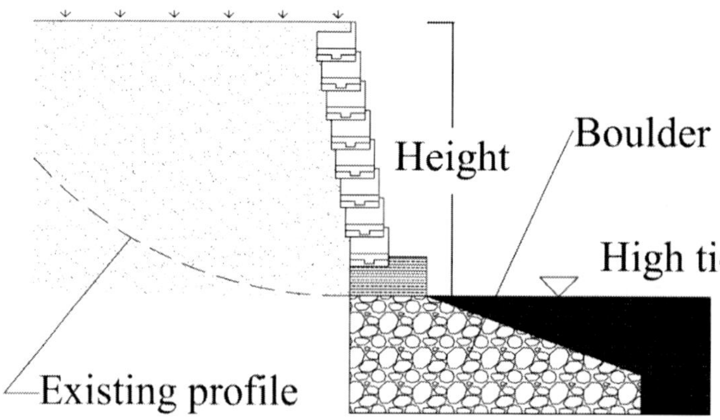

Figure 9.26. Proposed retaining wall using SIConSofa blocks for damaged road repair at sea side [courtesy, author].

9.6. Six years in Service of Drain cum Decorative Retaining Wall using SIConSofa Blocks

The proposed conventional RC retaining wall and RC drain for a project shown in Figure 9.27 was replaced using SIConSofa and precast RC bedding blocks as drain cum decorative retaining wall (Figure 9.28). The site work involvement in the conventional retaining wall and drain (Figure 9.27) needed several activities such as:

Activities for conventional RC Retaining wall:

1) Digging and removing of a large amount of back soil and foundation soil (Figure 9.27)
2) Fixation of formwork for casting of retaining wall
3) Placement of steel cage for retaining wall
4) Casting of concrete in-situ for retaining wall
5) Demoulding of formwork a few days later (as per specification)
6) Back fill and compaction

After completion of retaining wall, the activities for conventional RC Drain:

1) Alignment and fixation of inverted levels for drain
2) Fixation of formwork for casting of drain
3) Placement of steel cage for drain
4) Casting of in-situ concrete for drain
5) Demoulding of formwork a few days later (as per specification) and
6) Back fill and compaction for drain

The activities in the proposed new drain cum decorative retaining wall using SIConSofa and RC bedding blocks are:

1) Relatively less digging for foundation and back soil
2) Straight alignment and fixation of drain invert levels
3) Installation of RC bedding blocks in straight alignment and grooves interlocked
4) Installation of first layer with SIConSofa at both sides for load balance
5) Installation of second layer with SIConSofa blocks
6) Backfill and compaction after every two layers with SIConSofa installation

This new system using precast SIConSofa and RC bedding blocks (Figure 9.28) in installation took 1/3rd man-days as compared with the conventional drain and retaining wall shown in Figure 9.27. The use of precast SIConSofa and RC bedding blocks in the installation is very simple and rapid which saves installation time and workmanship.

9.6.1. Mass Production of SIConSofa and RC Bedding Blocks

Figure 9.29 shows the mass production and stacking of prototype SIConSofa blocks using normal weight concrete. The first batch of SIConSofa blocks was demoulded after $5^{1/2}$ hours of casting with the intention of optimising the mould facility for second time use on the same day. The second batch was demoulded the following morning and, the sequence of casting and demoulding were followed. The stacking was done after one day of casting.

Figures 9.30 and 9.31 show the precast RC bedding blocks. The same mould was used several times through careful handling. The grooves were formed using polystyrene blocks. The RC bedding blocks were cast in an inverted way and the casting face was placed onto the ground at site. Figure 9.31 shows RC bedding blocks ready for transportation to the site. A truck was used to carry all blocks from the production place to the construction site.

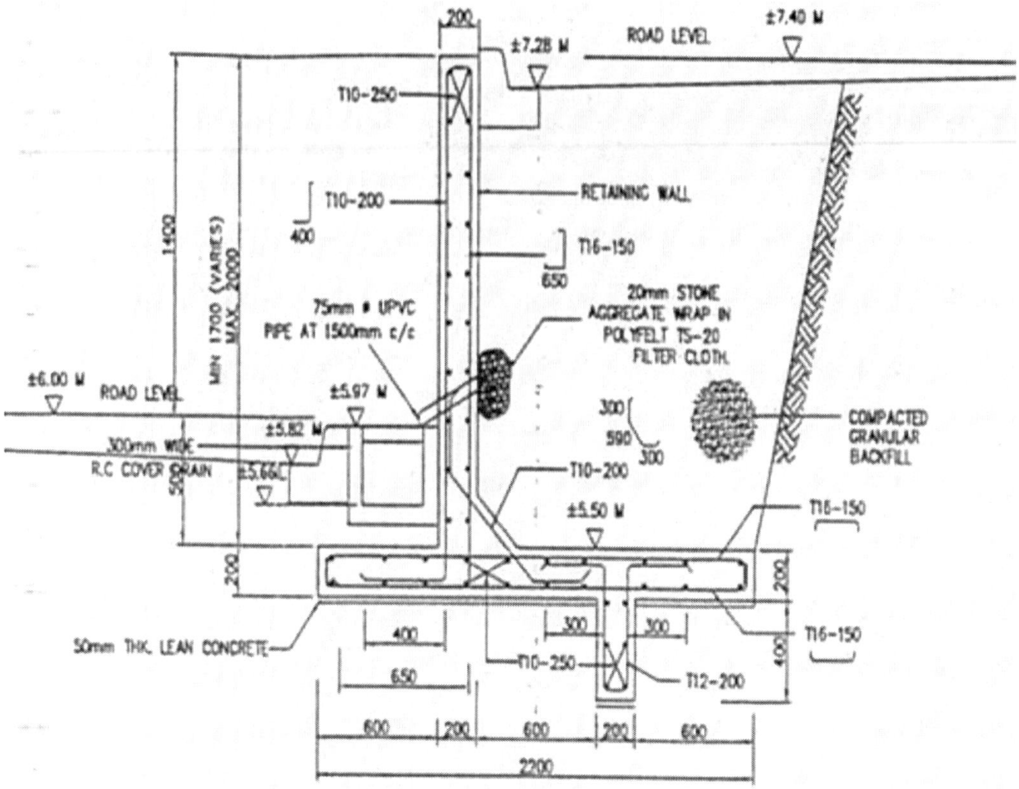

Figure 9.27. Conventional RC retaining wall and RC drain [courtesy: Bina Puri Constructions].

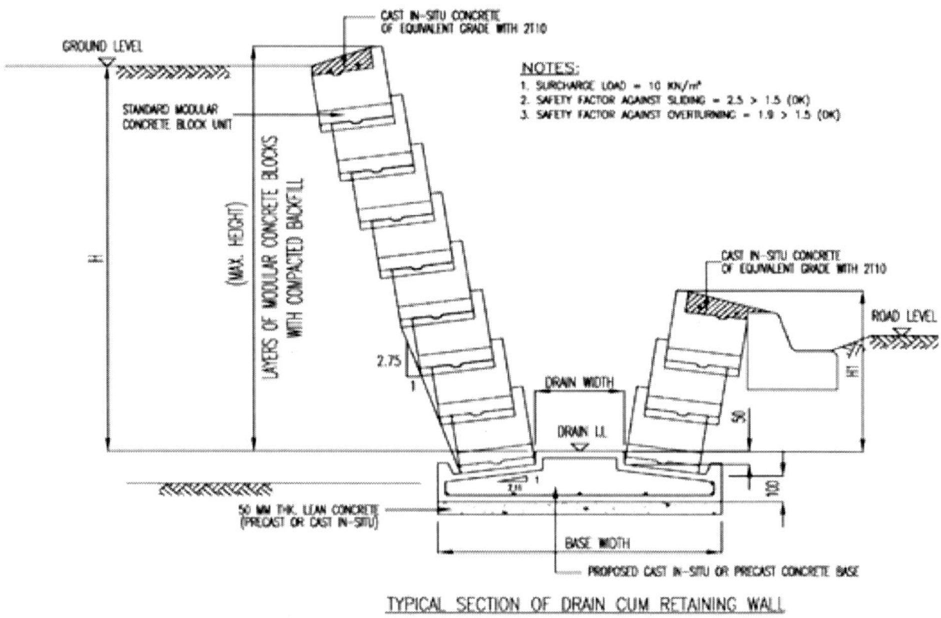

Figure 9.28. Drain cum decorative retaining wall using SIConSofa blocks [courtesy, Yong Lee Sun, personal contact].

Figure 9.29. Mass production and stacking of prototype SIConSofa blocks [courtesy, Hijau Ganda S/B].

Figure 9.30. Precast four units bedding blocks being demoulded manually [courtesy, Hijau Ganda S/B].

Figure 9.31. Precast RC bedding blocks ready for transportation [courtesy, Hijau Ganda S/B].

9.6.2. Installation of RC Bedding and SIConSofa Blocks at Site

This project was an L-pattern drain cum retaining wall. Figures 9.32 to 9.35 show the sequence of installation works of RC bedding and SIConSofa blocks at one side of the L-pattern.

The installation procedure was:

1) Fixation of straight alignment using thread and drain inverted levels at both ends
2) Installation of RC bedding blocks
3) Joining of grooves of RC bedding blocks using cast in-situ concrete with steel rods of 10mm dia. as nominal size
4) Installation of first layer of SIConSofa blocks at both sides of bedding blocks (with provision of extra space of about 300mm at the back side of SIConSofa layers as seen in Figure 9.32)
5) Backfilling and compaction done after installation of every two layers of SIConSofa blocks
6) Providing self-interlocking through installation of SIConSofa blocks
7) Use of Step 1 to 6 above for second part of the L-pattern
8) Use of brick wall to join both sides (Figures 9.36 to 9.38)
9) Fixation of Scupper drain with 150mm dia. PVC pipe (Figures 9.37 and 9.39)

Figure 9.39 shows the aesthetic appearance of drain cum retaining wall. The service age of this drain cum retaining wall is more than 6 years as seen in Figures 9.40 and 9.41. The plants grow on the wall surface (Figure 9.41) and it will cover the surface as a green retaining wall in urban condition.

Figure 9.32. Manual assembling sequence of precast RC bedding blocks followed by SIConSofa blocks with little gap provided at back of wall [courtesy, Bina Puri Constructions].

Figure 9.33. Manual assembling sequence of SIConSofa blocks followed by back filling [courtesy, Bina Puri Constructions].

Figure 9.34. Assembling of SIConSofa in progress followed by backfill compaction manually despite rain. Author's lovely PhD student, Elsa Eka Putri is seen observing the work progress [courtesy, Bina Puri Constructions].

Figure 9.35. Supervisors inspecting SIConSofa blocks as dry-stacked flexible structure for drain cum decorative retaining wall [courtesy, Bina Puri Constructions].

Figure 9.36. Assembled L-pattern drain cum decorative retaining wall about 2m high using self-interlocked SIConSofa precast blocks [courtesy, Bina Puri Constructions].

Figure 9.37. Fixation of 150mm dia. PVC scupper drain [courtesy, Bina Puri Constructions].

Figure 9.38. Corner joint using bricks for L-pattern drain cum decorative retaining wall [courtesy, Bina Puri Constructions].

Figure 9.39. L-pattern drain cum decorative retaining wall using SIConSofa blocks, retaining wall of different heights from 1m to about 2m high. Photo taken after completion of project [courtesy, Bina Puri Constructions].

Figure 9.40. Six years in excellent service of L-pattern drain cum decorative retaining wall using SIConSofa blocks, installation done in 2010 [courtesy, author].

Figure 9.41. Plant growing onto surface forming green landscaping on retaining wall [courtesy, author].

9.7. Slope Surface Protection Using SIConSofa Blocks

The widespread occurrence of landslides is associated with slopes and heavy rain. The construction of buildings, roads etc. across hilly areas, which involves cutting slopes to a certain extent causes the occurrence of landslides more frequently and visible.

On a soil slope, the gravity force of a solid object is derived from two force components. One component acts perpendicular to the slope and another acts parallel to slope [3]. The perpendicular force component acts to hold the solid object, whereas the parallel force component produces shear stress. This shear stress pulls the solid object down the slope. The resisting force known as shear strength which is a combination of frictional resistance and cohesion among soil particles such as clay, sand etc. that form the solid object. The solid object will move downward if the shear stress is larger than shear strength.

Rain water increases the weight on the soil slope as it seeps into the slope to replace the air from the pore space. So, the increased weight, especially due to rainstorms at the slope condition will increase the shear stress, leading to slope instability.

Generally, soil slope failure cases are seen due to the infiltration of water during rainstorms. The slope can collapse easily by the shear stress when the soil is in a nearly saturated condition and rapidly builds up pore pressure. Thus, during the rainstorm, the surface and ground water condition among other factors affect the stability of the slope.

Rainfall intensity and duration, slope angle and surface cover are among the major factors affecting the water infiltration characteristics of the unsaturated soil slope [4]. Generally, the slope is covered in several ways such as using vegetation surface protection and rigid surface protection. The Signal grass (Bracharia Decumben) and Mucuna bracteata (Legumes) act as vegetation surface protection and shotcrete or masonry acts as rigid surface protection, but they are seen ineffective in many cases (Figures 9.42 to 9.47).

The shotcrete, masonry or stone pitching is used as rigid surface protection (Figure 9.47). This system is highly impermeable. The weep holes are provided at regular intervals, horizontally and vertically to reduce the build up of pressure on the slope during rain. Sometimes soil nailing is also done including shotcrete application. The hydroseeding method involves using grass seed sprayed onto pretreated surfaces with fertilizer in moist conditions (Figures 9.42 and 9.43), so grass can grow quickly. The grass root will act as a reinforcing fibre to hold surface soil. Similarly, the Mucuna bracteata plant is also used for slope surface protection. (Figure 9.45 and 9.46).

The slope failure during heavy rain can be minimized if the slope is covered in a proper manner which will not allow water to penetrate during rain but allows internal water (water in hill) to flow out from the bottom of the slope. Thus, if the slope surface is covered like fish-scales using SIConSofa blocks, it will not allow rain to penetrate into the soil. The conceptual approach for slope surface protection using SIConSofa blocks is shown in Figures 9.48 to 9.56. Hence, it needs further investigation to fulfil the requirements as per the construction industry.

Figure 9.42. Conventional slope surface protection using grass through bed preparation with grass seed spraying, [courtesy, Hijau Ganda S/B].

Figure 9.43. Conventional slope surface protection covered with green grass [courtesy, Hijau Ganda S/B].

Figure 9.44. Slope surface failure even though covered with grass, [courtesy, GCU Consultants S/B].

Figure 9.45. Conventional slope surface protection using the planting of Mucuna Bracteata [courtesy, Hijau Ganda S/B].

Figure 9.46. Rapid wide spreading of the Mucuna Bracteata plant, even covering the roadside drain [courtesy, Hijau Ganda S/B].

Figure 9.47. Slope failure built with conventional practice [courtesy, GCU consultants S/B].

Figure 9.48. Front view of slope surface assembled inversely using SIConSofa model blocks, appearance as fish scales [courtesy, author].

Figure 9.49. Side view of slope surface assembled inversely using SIConSofa model blocks [courtesy, author].

Figure 9.50. Inverted assembly of prototype SIConSofa blocks for slope surface covering, a mock covering [courtesy, author].

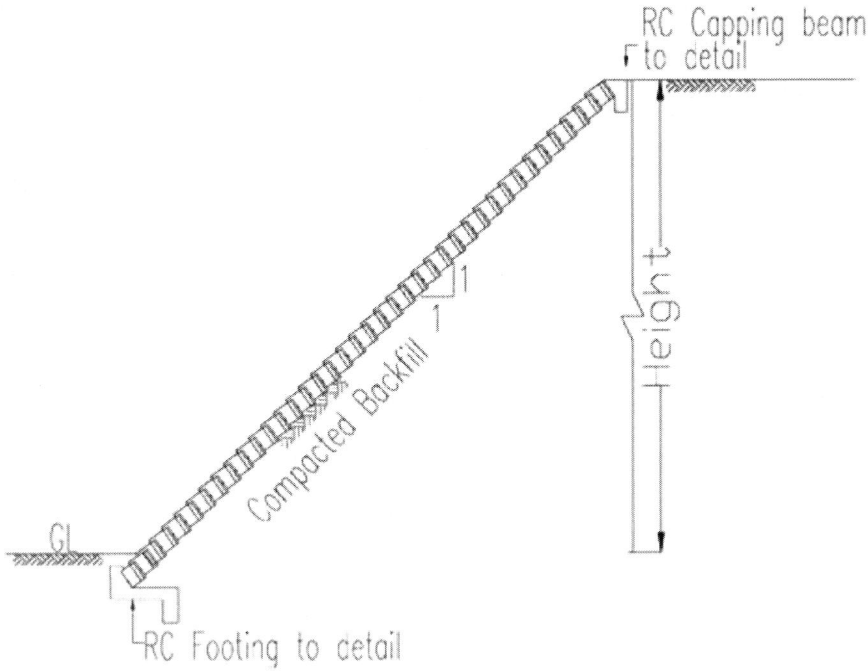

Figure 9.51. Proposed surface covering with inverted assembly of prototype SIConSofa blocks [courtesy, author].

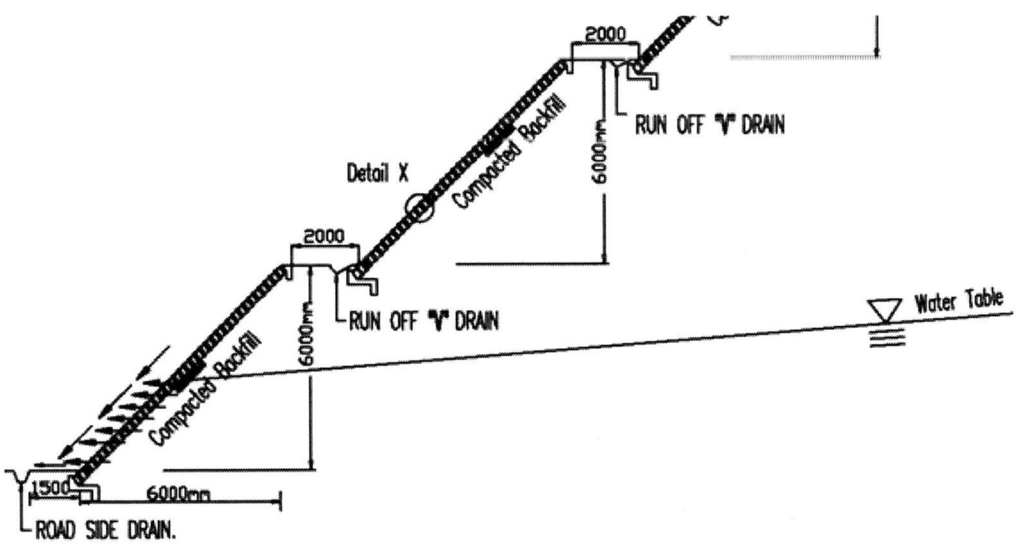

Figure 9.52. Proposed typical slope surface covering with inverted assembly of prototype SIConSofa blocks for several slopes [courtesy, author].

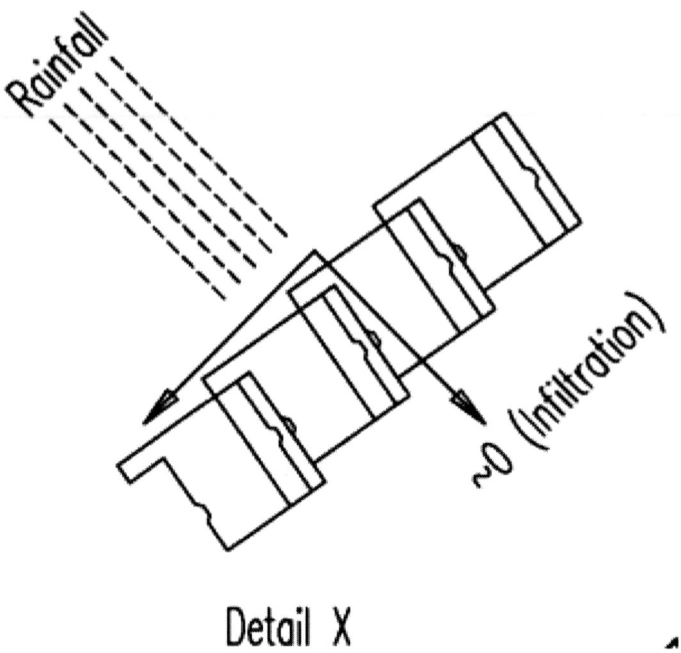

Figure 9.53. Detail X referring to Figure 9.52 above [courtesy, author].

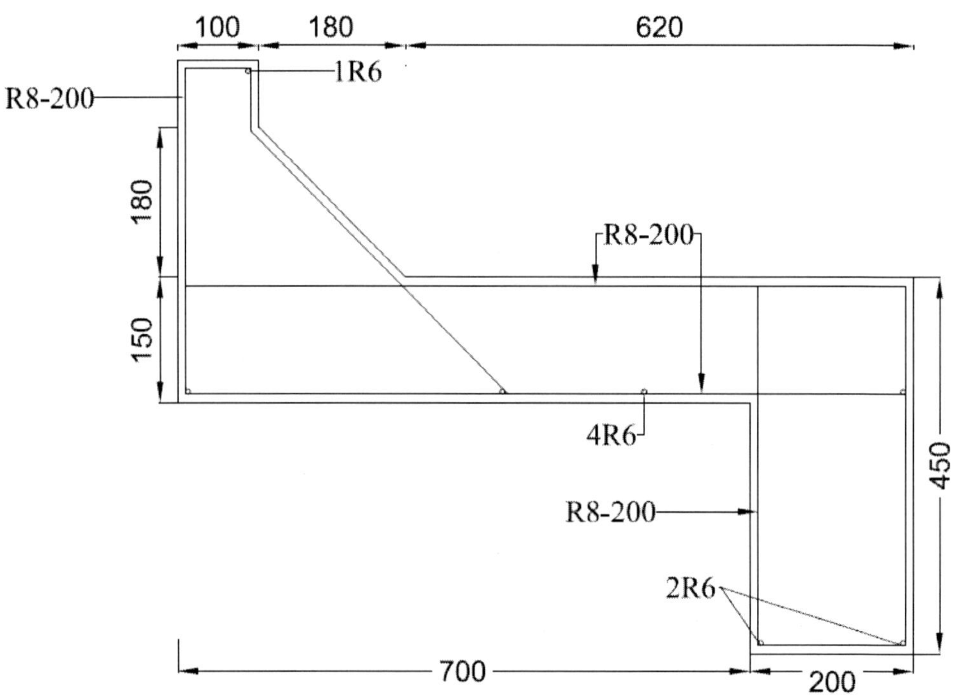

Figure 9.54. Details of precast RC Toe of 600mm long block, referring to Figures 9.51 and 9.52 above [courtesy, author].

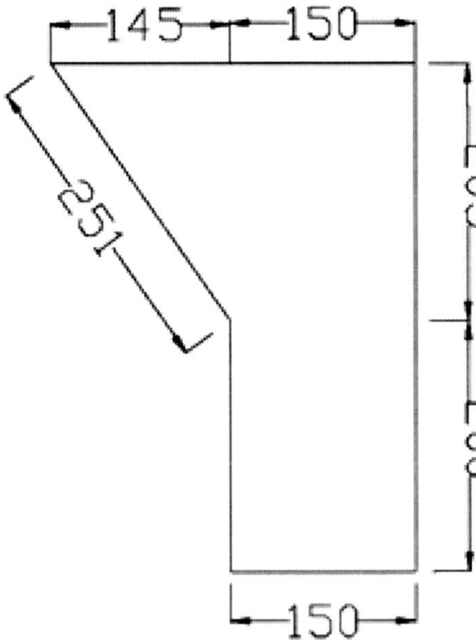

Figure 9.55. Typical shape of Capping beam referring to in Figures 9.51 and 9.52 above, with provision of nominal reinforcement [courtesy, author].

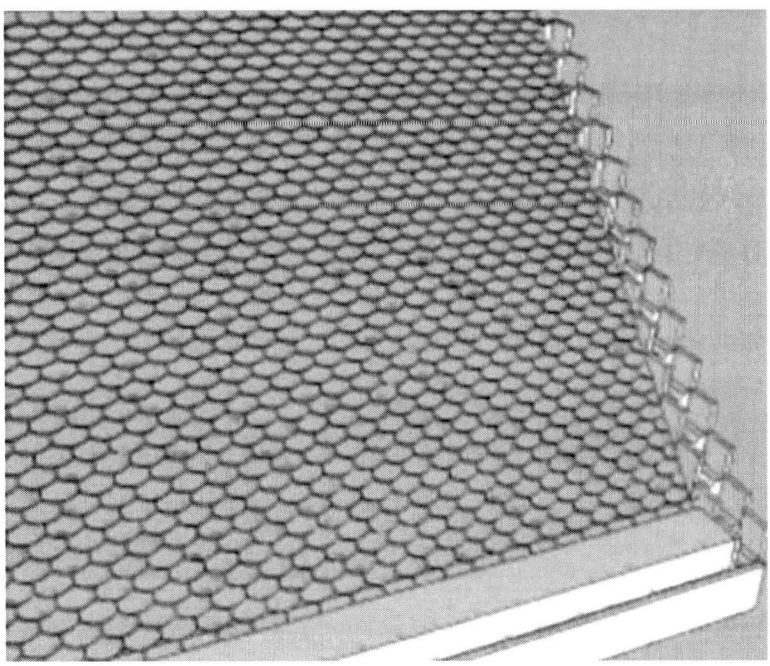

Figure 9.56. Proposed typical front view of slope surface built using SIConSofa blocks with cascaded drain at side [courtesy, author].

9.8. Chapter Remarks

The precast SIConSofa blocks satisfy the requirement of retaining wall structures. This environmental-friendly SIConSofa has features of modularity, adaptability, portability and is self-interlocking. It absorbs minor movement due to the dry-stacked flexible structure. The installation is very simple and rapid which saves installation time and workmanship. This precast block is the potential alternative to the conventional system for the drain, retaining wall and slope surface protection.

References

[1] Research Report. 2006. Performance of concrete made from agricultural and industrial solid wastes and by-products, *IRPA research grant no. 03-02-10-0033-EA0031,* Part-3. Precast blocks for retaining wall. Universiti Malaysia Sabah.

[2] N. S. V. Kameswara Rao, *Foundation Design: Theory and Practic*e, Publisher: Wiley, ISBN: 978-0-470-82534-1.

[3] Prof. Stephen A. Nelson. Slope Stability, Triggering Events, Mass Movement Hazards, Natural Disasters. *Tulane University*. Retrieved on 6 April 2016.

[4] Bujang B. K. Huat, Faisal HJ. Ali, T. H. Low. 2006. Water infiltration characteristics of unsaturated soil slope and its effect on suction and stability, *Geotechnical & Geological Engineering*. 2006. 24(5). pp 1293-1306.

Chapter 10

IBS Concrete Products for Water Village and Eco-Resort in a Marine Environment

Keywords: Marine environment, Water village, Eco-resort, IBS concrete products, C-channel, Beam, Single round pile cap, round pile

10.1. Introduction

A water village is built in a water body like seas and rivers. In Brunei, Kampong Ayer (Kampung Air, water village) retains much of the historical features of the 16th century. This water village is made up of 42 small villages linked together by more than 29,140 meters of walkways and over 4,200 structures including homes, mosques, restaurants, shops, schools, and a hospital for 30,000 people. They are stilt houses and wooden walkways. The lifestyle of this community is as river traders, fishermen and sellers of traditional handicrafts [1].

In Sabah, Malaysia, water village settlements have been known as the common traditional settlements along most of the coast, island, and estuaries. In particular, the water village (Figures 10.1 and 10.2) existed in Malaysia before the 19th century.

Ecotourism is defined by the International Ecotourism Society, that responsible travel to natural areas where they typically have a special natural environment that improves the welfare of local people. Several factors under ecotourism are (i) conserve and enhance the environment, (ii) educate and inform the visitors, (iii) support conservation and community-based activities, and (iv) sustain the social, cultural and financial well-being of local residents, small businesses and communities [2].

Marine tourism includes the recreational activities being held at places of a marine environment such as scuba diving and snorkeling, wind surfing, fishing, sea kayaking, etc. The marine environment is known to be with saline and tide affected water.

A marine resort is generally built with natural surroundings without disturbing the serene environment. It is an ideal place for people who can spend time happily with family, friends for various recreational activities for a short duration.

Water supply and sewerage systems are crucial for human health. One common issue is unhealthy sanitation systems exit in such water villages resulting in pollution of the water body due to the direct discharge of human excreta. Traditional materials like wood, metal sheets etc are used to build these houses and walkways without engineering input. The water-village houses being remote communities and detached from land can be improved with modern infrastructure systems for better living standards of inhabitants through IBS incorporating water and sewerage facilities. The eco-friendly tourist destinations such as island hotels, eco-friendly resorts, family-friendly recreation parks, water-theme parks etc. can be developed.

The protection of the natural environment and marine reserves is a key priority for the marine ecosystem due to environmental conservation, tourism, and economic gain. The pristine beauty of the marine ecosystem must be protected and treated so that its beauty can be enjoyed as a sustainable marine ecosystem, especially as a tourism destination.

Figure 10.1. Crystal clear water - an ideal place for an eco-tourism resort. Water village can be seen in the distance. [courtesy, author].

Figure 10.2. Crystal clear water being polluted by direct discharge of human excreta etc. at water village [courtesy, Cassondraizzatiarby, personal contact].

10.2. CONSTRUCTION DIFFICULTIES IN A PRISTINE, BEAUTIFUL MARINE ENVIRONMENT

Figure 10.1 shows crystal clear water which is an ideal place for the construction of an eco-tourism resort. Under such conditions, the spine is needed for the connection between the beach and villas etc. to allow for the movement of people without contact with water. Since the spine is built on water, several problems will occur due to the surrounding conditions and the environment issue.

One of the major construction problems in such areas is the storage of construction materials. It is difficult to store large amount of materials on the sea due to several factors such as corrosion of reinforcing bars, shallow water etc. Therefore, all steel bars should be stored far away from the seawater. This will cause an inconvenience to the construction if the workers can move only a certain amount of steel bars required to construct a certain part. In addition, it is difficult to provide a proper working platform on the sea water to allow the workers to do the construction and store all materials and rubbish.

Another issue is to cast concrete beneath the sea water. The foundations of houses and spines is beneath the water since they are built above the sea water. A pile foundation is the only option for the construction of the spines and houses because the piles can act as columns to support the structures.

The delivery of construction materials is also a major problem in such marine projects. Since the construction site near the beach has shallow water, it is difficult to use a large boat. The required materials can only be delivered in small amounts using small boats. Furthermore, the conventional curing process using fresh water is not possible in such areas.

Since the spine, villas etc. are built on such areas, platforms should be provided for the workers to do the construction. But it is difficult to provide a platform on the sea water. Therefore, constructing such a project with full precast structural components will ease the contractor's construction work. Since there is no formwork and rebar is required while using precast structural components, no platform is required to be built on the sea.

Another important issue in pristine, beautiful marine areas is the impact of construction to the environment. The pristine beauty of the marine area is the main attraction to tourists due to beautiful underwater view and the cleanliness of the sea water. Rubbish pollutes the sea water around the construction area. Despite the cleanliness of the sea water, the rubbish may also destroy the underwater ecosystem. In addition, concrete waste may drop into the sea during concreting or the transportation of the concrete which may cause pollution to the seawater.

10.3. SOLAR-POWERED WATER PURIFIER

The solar-powered water purification system, promises to help change the lives of rural and remote communities by providing safe and clean water using renewable and sustainable energy. There are several filters to purify the water from harmful contaminants such as pesticides, bacteria, chemicals, dirt, and other pollutants. In addition, it is also connected to a solar panel which is installed on the roof to power the battery that supplies electricity for lighting and other purposes in the house. The system can filter and purify up to seven litres per minute of raw water from any source including rain and rivers. The system provides a cost-effective and sustainable solution of potable water for rural and remote communities [3].

10.4. Bioreactor for Sewage Treatment

The BioGill bioreactor has been used for the Mantaray Island resort, Fiji which is too far from the mainland to treat sewage, commercial kitchen and grease trap wastewaters since 2013 [4]. The treatment system is easy to install, operate and maintain. The modular design of the bioreactors has also made transportation and installation easier for the resort management. The system is designed to treat up to 15 m^3/day of influent which is more than a 65 population equivalent (PE). The daily wastewater discharge is 225 liters/person. The system with high-load sewage, rich in oils from a commercial kitchen with an average BOD of 662 ppm reduces to an average BOD of 25 ppm (96% removal efficiency) in a 24 hour cycle. Chemical Oxygen Demand (COD) of 998 ppm is reduced to 58 ppm, a 94% reduction. Total nitrogen is reduced on average from 112 ppm to only 24 ppm (79% efficiency). The system generates wastewater for subsurface irrigation for the resort's gardens that have grown into a beautiful, lush jungle. The island, through BioGill technology is one of the most environmentally friendly resorts in Fiji. Therefore, such a system for sewage treatment plays a vital role in remote areas. However, this chapter gives focus mainly on the construction of a walkway or spine, or house using IBS precast structural components.

10.5. IBS Precast Concrete Products

The conventional method with cast in-situ concrete is not user and eco-friendly while constructing walkways, houses etc. in the water especially for the marine ecosystem. Precast concrete in another way is beneficial due to its high quality and other benefits. There are several advantages of using IBS precast products as discussed in Section 7.5.1 (Chapter 7).

It is suitable to use precast components in marine construction as it minimizes the pollution to the environment during construction. Pile foundations in water act as columns to support the house and walkway. It also can overcome formwork issues if structural components are precast in such a situation.

The connection design is important in precast components because the structural capacity greatly depends on their connection with other members. Furthermore, the sealing method for the gaps in the connection must also be considered. Displacement is expected due to several external actions like waves, currents, wind etc. so tolerances are important for precast components particularly structures in water. A precise surveying is required to be carried out under such circumstances.

10.6. Structural Elements in Floors and Walkways

The main structural elements for projects such as houses and walkways are precast C-channels such as floor panels, beams, round pile caps and piles.

10.6.1. Precast C-Channel Floors

The precast C-channel as shown in Section 7.5.4 (Chapter 7) has unique benefits in terms of less concrete usage resulting in reducing the dead load for the foundation. It is developed with 1m width and varying spans from 3 to 8m. It shows the highest structural performance both at ultimate and service conditions. This system does not need topping concrete once assembled at the final site. The provision of service facility installation is proposed at the soffit area with a maximum vertical open space of 275mm. This vertical open space is suitable to accommodate pipes for water and sewerage systems. This C-channel is cost effective and offers a simple way of production, transportation, erection and installation. Thus it is recommended for use in house and walkway construction particularly in eco-friendly resorts in marine conditions.

10.6.2. Precast RC Beam

A typical RC beam for 4m wide spine is proposed using a precast unit. The details of such an RC beam analysis and design using Eurocode 2 are shown below.

Specification:

Effective span, $L = 2\ m$
Characteristic Actions:
Permanent, $g_k = 16.23\ kN/m$ (including self-weight)
Variable, $q_k = 9\ kN/m$
Design life = 50 years (Table 2.1 EN 1990)
Fire Resistance = R60 (Sec. EN 1992-1-2)
Exposure classes = XS 1

Materials:

Characteristic strength of concrete, $f_{ck} = 30\ N/mm^2$ (Table F.1 EN 206)
Characteristic strength of steel, $f_{yk} = 500\ N/mm^2$
Characteristic strength of link, $f_{yk} = 500\ N/mm^2$
Unit weight of reinforced concrete = $25\ kN/m^3$ (Table A.1 EN 1991-1)
Assumed:
$\emptyset_{bar\ 1} = 16\ mm$
$\emptyset_{link} = 10\ mm$

RC beam size:

Overall depth, $h = \frac{L}{13} = \frac{2000}{13} = 153 \cong 350$
Width, $b = 0.4\,h = 0.4 \times 153 = 61 \cong 500$
Use:
$b \times h = 500 \times 350$

Durability, Fire and Bond Requirements:

Min. cover with regard to bond, $C_{\min b} = 20\ mm$
Min. cover with regard to durability, $C_{\min dur} = 40\ mm$

Main Reinforcement:

Effective depth, $d = h - C_{nom} - \emptyset_{link} - 0.5\ \emptyset_{bar} = 292\ mm$

Design bending moment, $M_{Ed} = 31.31\ kNm$ [from beam analysis]

$$K = \frac{M}{bd^2 f_{ck}} \leq 0.167$$

$$= \frac{31.31 \times 10^6}{500 \times 292^2 \times 30}$$

$$= 0.024 < 0.167$$

$K < K_{bal}$

Compression reinforcement is not required.

$$z = d\left[0.5 + \sqrt{0.25 - \frac{K_{bal}}{1.134}}\right] \leq 0.95\ d$$

$$= 0.98\ d$$

$$z = 0.95 \times 292 = 277.4\ mm$$

Area of tension steel,

$$A_s = \frac{M}{0.87\ f_{yk}\ z} = \frac{31.31 \times 10^6}{0.87 \times 500 \times 277.4}$$

$$= 257.56\ mm^2$$

$$\cong 258\ mm^2$$

Minimum and maximum reinforcement area,

$$A_{s\,min} = 0.26\left(\frac{f_{ctm}}{f_{yk}}\right) bd;$$

where, $f_{ctm} = 0.3(f_{ck})^{\frac{2}{3}}$ for $f_{ck} \leq 50$

$= 0.3\ (30)^{\frac{2}{3}} = 2.93$

$$= 0.26\left(\frac{2.93}{500}\right) bd$$

$$= 0.0011\ bd \geq 0.0013\ bd$$

$$= 0.0013 \times 500 \times 292 = 222\ mm^2$$

$$A_{s\,max} = 0.04\ A_c = 0.04\ bh$$

$$= 0.04 \times 500 \times 350 = 7000\ mm^2$$

Hence, Main steel provided with 3H 16 for both top and bottom faces

Shear Reinforcement:

Design shear force, $V_{Ed} = 65.14\ kN$ [from beam analysis]

Concrete strut capacity,

$$V_{Rd\,max} = \frac{0.36\, b_w d f_{ck}\left(1 - \frac{f_{ck}}{250}\right)}{(cot\,\theta + \tan\theta)}$$

$$= \frac{0.36 \times 500 \times 292 \times 30 \times \left(1 - \frac{30}{250}\right)}{(cot\,\theta + \tan\theta)}$$

= 481 kN; where, θ = 22° cot θ = 2.5

= 693 kN; where, θ = 45° cot θ = 1.0

$V_{Ed} < V_{Rd\,max}$ $\theta = 22°$

$V_{Ed} < V_{Rd\,max}$ $\theta = 45°$

Therefore angle θ < 22°

$$\theta = 0.5\, sin^{-1}\left[\frac{V_{Ed}}{0.18\, b_w d f_{ck}\left(1 - \frac{f_{ck}}{250}\right)}\right]$$

$$= 0.5\, sin^{-1}\left[\frac{65.61 \times 10^3}{0.18 \times 500 \times 292 \times 30 \times \left(1 - \frac{30}{250}\right)}\right]$$

$$= 0.5\, sin^{-1}\{0.09\} = 2.69\,^{\circ}$$

Use: $\theta = 22.0\,^{\circ}$ $\tan\theta = 0.40$ $cot\,\theta = 2.48$

Shear links,

$$\frac{A_{sw}}{s} = \frac{V_{Ed}}{0.78\, f_{yk}\, d \cot\theta}$$

$$= \frac{65.61 \times 10^3}{0.78 \times 500 \times 292 \times 2.48}$$

$$= 0.23$$

Try link: H10 $A_{sw} = 157\, mm^2$

Spacing, s

$$s = \frac{157}{0.23}$$

$$= 682\, mm$$

Max. spacing, $S_{max} = 0.75\, d = 0.75 \times 292 = 219\, mm$

Use H10 – 200

Minimum links:

$$\frac{A_{sw}}{s} = \frac{0.08\, f_{ck}^{\frac{1}{2}}\, b_w}{f_{yk}}$$

$$= \frac{0.08 \times (30)^{\frac{1}{2}} \times 500}{500}$$

$= 0.438$

Try link: H10 $A_{sw} = 157\ mm^2$

Spacing, s

$$s = \frac{157}{0.438}$$

$= 358$ mm $\leq 0.75\,d$

Use H10 – 200

Shear resistance of minimum links,

$$V_{min} = \left(\frac{A_{sw}}{s}\right)(0.78\,d\,f_{yk}\cot\theta)$$

$= 221.7\ kN$

Additional longitudinal reinforcement:

Additional tensile force,

$\Delta F_{td} = 0.5\,V_{Ed}\cot\theta$ $\leq \left(\frac{M_{Ed\,max}}{z}\right)$

$= 0.5 \times 65.61 \times 2.48 = 80.77\ kN$

$$\left(\frac{M_{Ed\,max}}{z}\right) = \frac{31.31 \times 10^6}{277.4}$$

$= 112\ kN$

Additional longitudinal reinforcement,

$$A_s = \frac{\Delta F_{td}}{0.87 f_{yk}}$$

$$= \frac{80.77 \times 10^3}{0.87 \times 500}$$

$= 186\ mm^2$ [To be added to the A_s near support (after curtailment)]

Deflection:

Percentage of required tension reinforcement,

$$\rho = \frac{A_{s\,req}}{b\,d} = \frac{258}{500} \times 292 = 0.0017$$

Reference reinforcement ratio,

$$\rho_0 = (f_{ck})^{\frac{1}{2}} \times 10^{-3} = (30)^{\frac{1}{2}} \times 10^{-3} = 0.0055$$

Factor for structural system, K = 1.0

$$\frac{l}{d} = k\left[11 + 1.5\sqrt{f_{ck}}\,\frac{e_0}{e} + 3.2\sqrt{f_{ck}}\left(\frac{e_0}{e} - 1\right)^{\frac{3}{2}}\right] \qquad \text{if, } e \le e_0$$

$$= 1.0\left[11 + 1.5\sqrt{30}\,\frac{0.0055}{0.0017} + 3.2\sqrt{30}\left(\frac{0.0055}{0.0017} - 1\right)^{\frac{3}{2}}\right]$$

$$= 1.0\,[11 + 26.56 + 17.52\,(3.34)]$$

$$= 96.07$$

Modification factor =1 for span 2 m <7m span)

Modification factor for steel area provided,

$$= \frac{A_{s\,prov}}{A_{s\,req}} = \frac{258}{603} = 0.43 \le 1.5$$

Therefore allowable span-effective depth ratio,

$$\left(\frac{l}{d}\right)_{allowable} = 96.07 \times 1 \times 0.43 = 41.31$$

Actual span-effective depth,

$$\left(\frac{l}{d}\right)_{actual} = \frac{2000}{292} = 6.85 < \left(\frac{l}{d}\right)_{allowable}$$

Cracking:

Limiting crack width, $w_{max} = 0.3\,mm$

Steel stress,

$$f_s = \frac{f_{yk}}{1.15} \times \frac{G_k + 0.3\,Q_k}{1.35\,G_k + 1.5\,Q_k}\,\frac{A_{s\,req}}{A_{s\,prov}}\,\frac{1}{\delta}$$

$$= \left(\frac{500}{1.15}\right) \times \left[\frac{(16.23 + (0.3 \times 9.0))}{35.41}\right] \times \left(\frac{258}{603}\right) \times 1.0$$

$$= 99.38\,\frac{N}{mm^2}$$

Max. allowable bar spacing = 300 mm

Bar spacing,

$$s = \frac{[500 - 2(40) - 2(10) - (16)]}{2}$$

$$= 192 < 200\,mm$$

Hence, Stirrup provided with H10-200

The cross section of an RC beam with reinforcement details and typical beam pattern are shown in Figures 10.3 and 10.4 respectively.

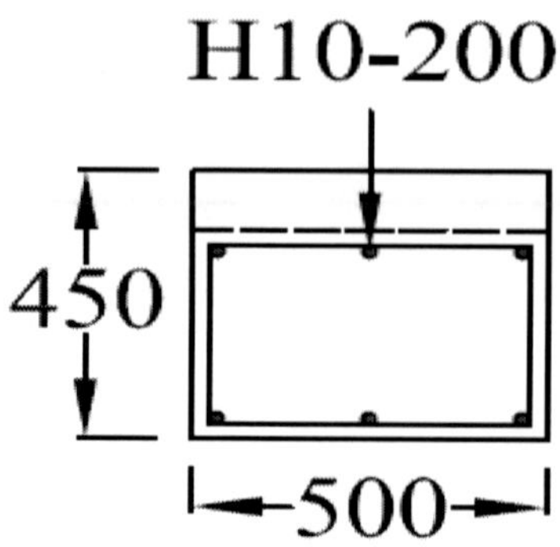

Figure 10.3. Cross section (500mm wide and 350mm thick, and end thickness of 450mm) of precast RC beam [courtesy, author].

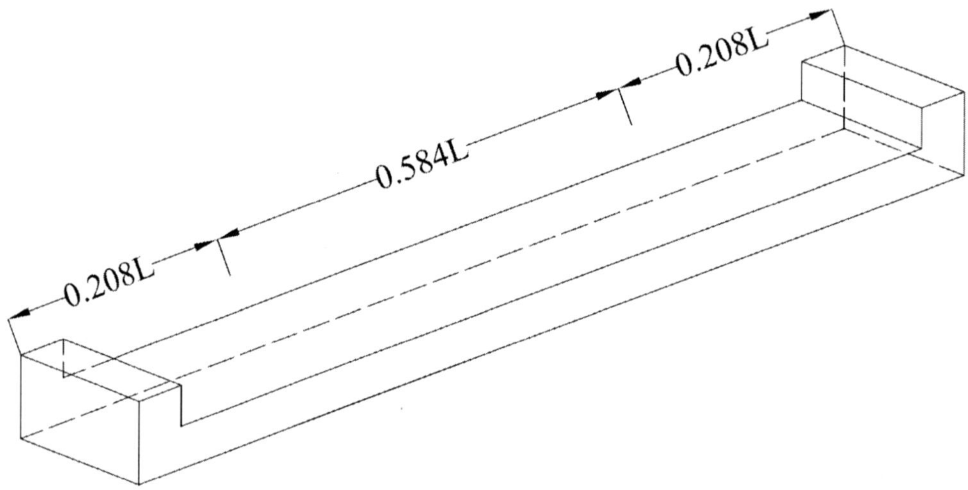

Figure 10.4. Typical pattern RC beam including location of lifting hooks, lifting hooks fixed at 0.208L from each end for up to 5m long beam [courtesy, author].

10.6.3. Set up of Precast RC Components

Figures 10.5 to 10.8 show typical arrangement of precast C-channels onto the precast beam, followed by precast beam onto precast round pile caps, and finally pile caps onto precast round piles. The required tolerances are provided for each setting. The gaps are required to be filled using grout. The provision of service pipes such as water and sewer are also shown in Figure 10.6. These types of precast units can overcome the difficulties as mentioned in Section 10.2.

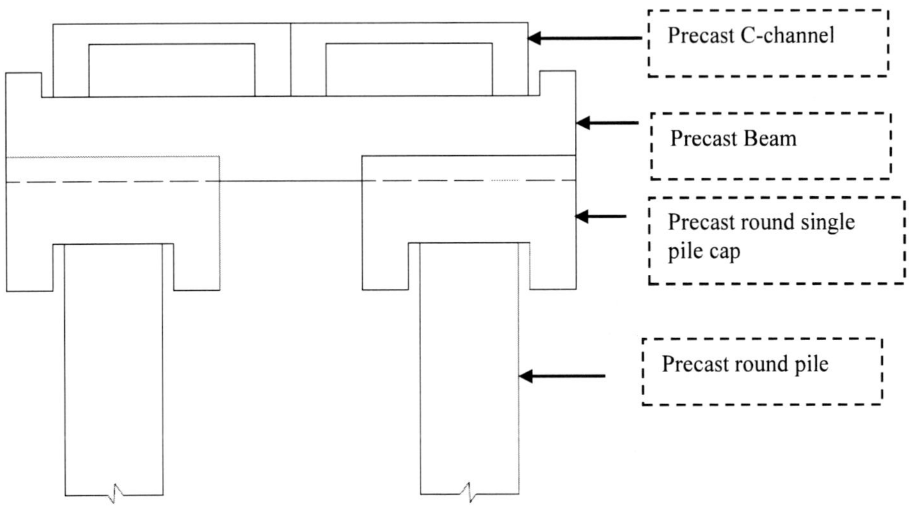

Figure 10.5. Typical crosss section of walkway using two C-channels assembled onto precast components [courtesy, author].

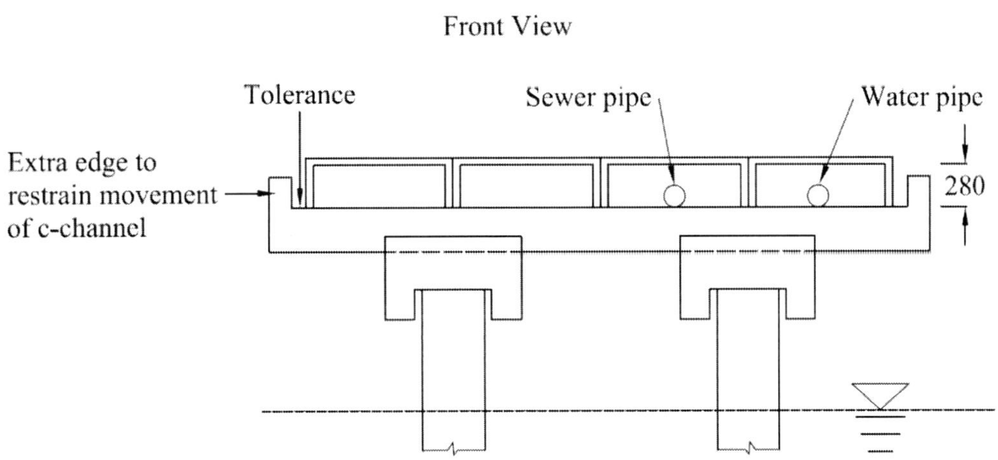

Figure 10.6. Typical crosss section of four C-channels assembled onto precast beam [courtesy, author].

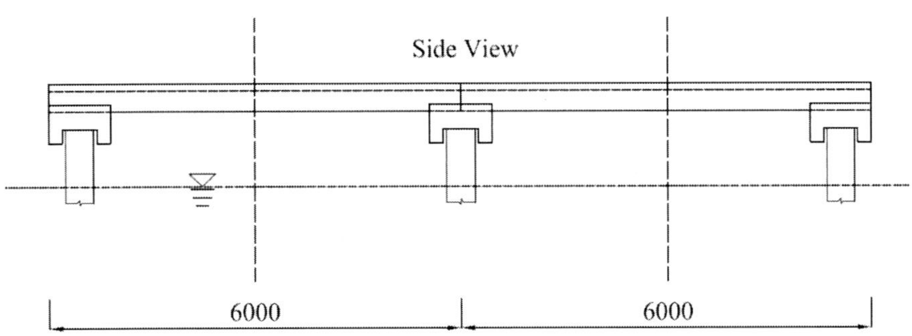

Figure 10.7. Typical longitudinal section with 6m span using C-channels assembled onto precast beams [courtesy, author].

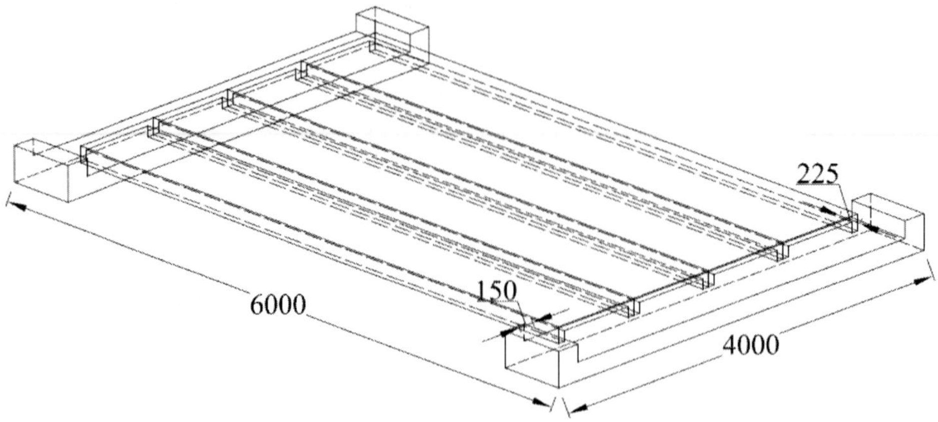

Figure 10.8. Typical perspective view of C-channel platform assembled onto precast beams [5].

10.6.4. Precast Round Single Pile Cap

It is assumed that the precast pile cap is a single unit with circular shape. The diameter is chosen based on the width of the precast beam and the depth is chosen based on structural and stability requirements with precast piles. A spun pile of 350mm diameter is proposed. The shape and dimensions are shown in Figures 10.9 to 10.12.

A little gap is provided for proper adjustment of the beam with the pile cap and also for grouting. For proper fixation of a pile cap with a pile, a circular open space with 300mm deep and 400mm diameter is formed at the bottom face of the pile cap. The 350mm diameter pile is selected with 300mm embedded in the pile cap. The remaining open space after fixation is filled with grout.

Design parameters:

Cantilever beam action (monolithic behaviour between cap and pile)
Eccentricity=50mm
Moment =65.14kN×50mm=3.26 kNm
f_{cu}=40MPa (28-day comp. strength of concrete)
Breadth, b=1000mm
Effective depth, d= 300-50=250mm

$K=[M/\{f_{cu}bd^2\}]<0.167$
$=[3.26\times10^6/\{40\times1000\times(250)^2\}]$
$=0.0013$
$z=d[0.5+\{0.25-(k/1.134)\}^{1/2}\leq0.95d$
Use, z=0.95d=237.5mm

$A_{s\ required}=[M/(0.87f_yz)]$

where,

f_y=500MPa
$A_{s\ required}$= 31.55mm^2/m

$A_{s\ minimum}$= 0.13bh%=0.13×1000×300=390mm^2/m

Hence, $A_{s\ provided}$ = H12-250mm (452mm^2/m) top and bottom layers for both ways

Lifting Hooks

According to Eurocode 2 for anchorage length,

$l_{bd} = \alpha_1\alpha_2[f_{yk}/4.6f_{bd}]\emptyset$

where,

l_{bd} - required anchorage length

α_1- bar is bent into round pile cap and concrete cover coefficient, c_d is equivalent to 4 times of bar diameter (4$\emptyset$), and is larger than 3$\emptyset$, so value of α_1 = 0.7

$\alpha_2 = 1\text{-}0.15(c_d\text{-}3\emptyset)/\emptyset = 1\text{-}0.15(4\emptyset\text{-}3\emptyset)/\emptyset = 0.85$

f_{yk} = 500 MPa

f_{bd} = 2.1 MPa,

$\emptyset$ - diameter of bar

Hence, l_{bd} = 31$\emptyset$, [use 12mm diameter bar]

So, anchorage length is 31×12mm =372mm = 400mm

Hence, Provide 2 nos. hooks using H12 with each leg of 400mm long.

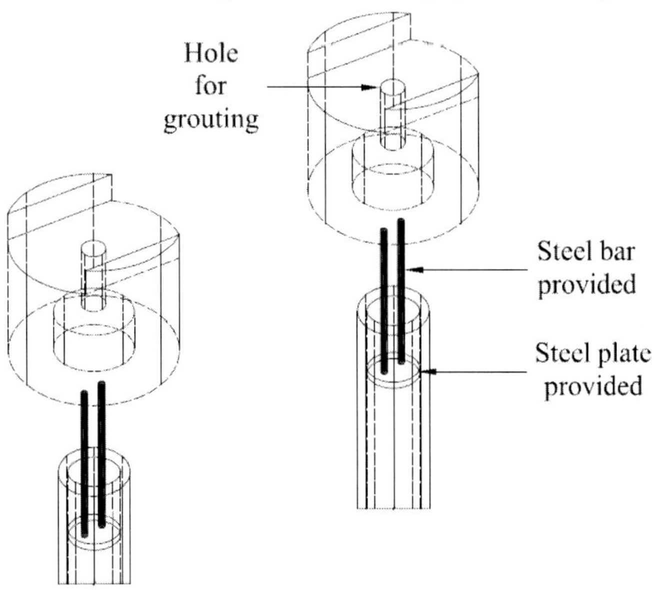

Figure 10.9. Typical perspective view of precast single unit pile cap being assembled onto circular precast pile [5].

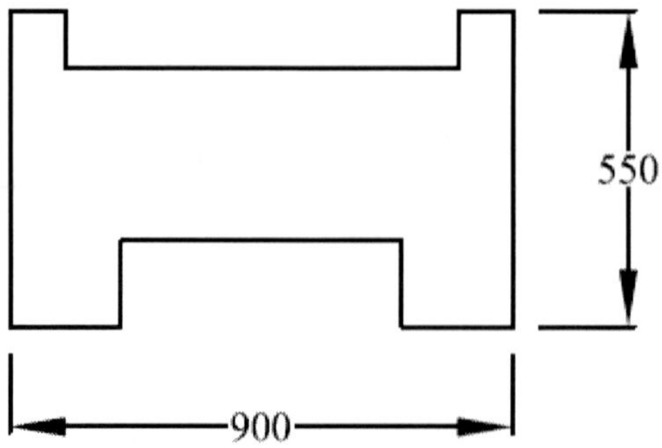

Figure 10.10. Two dimensional view of circular single unit precast pile cap [courtesy, author].

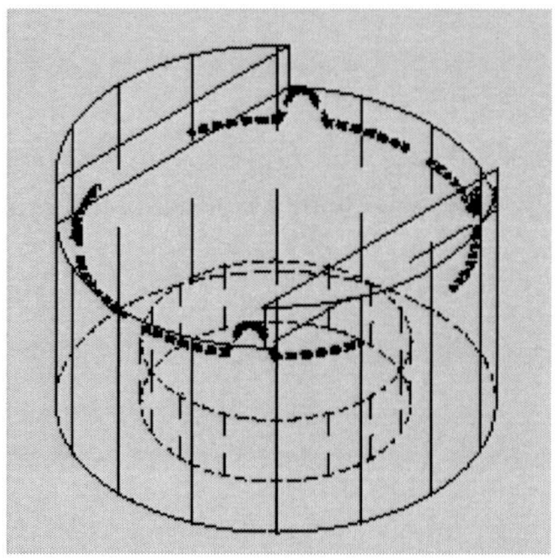

Figure 10.11. Lifting hook set up for single circular pile cap [5].

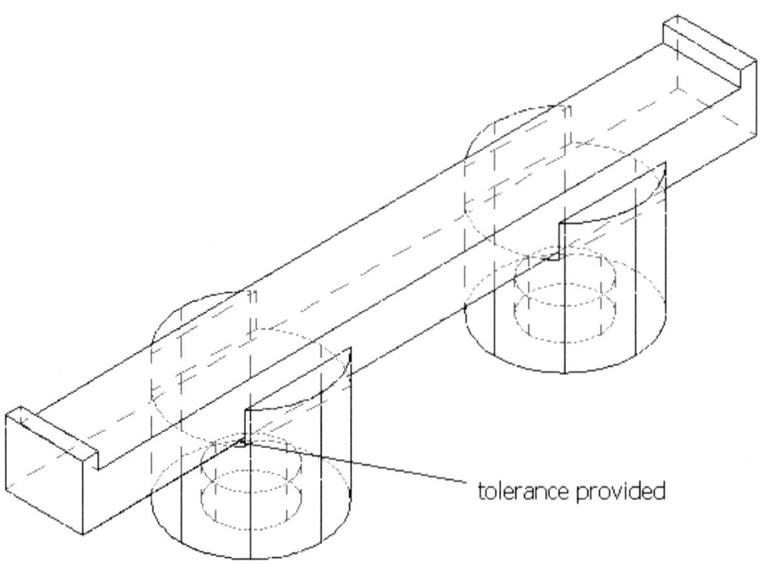

Figure 10.12. Conceptual view of assembled precast beam onto precast round pile caps [5].

10.4.5. Conceptual Plan of Platform Using Precast Units

Figure 10.13 shows the platform using model blocks of piles, pile caps, beams and C-channels. Figures 10.14 and 10.15 show conceptual plan of house, spine and villa.

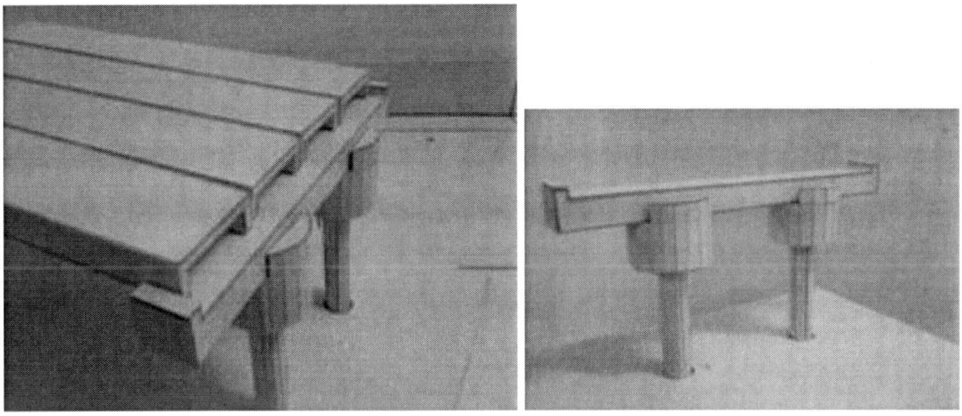

Figure 10.13. Physical platform assembled using model blocks of C-channels, beams, circular pile caps and piles [5].

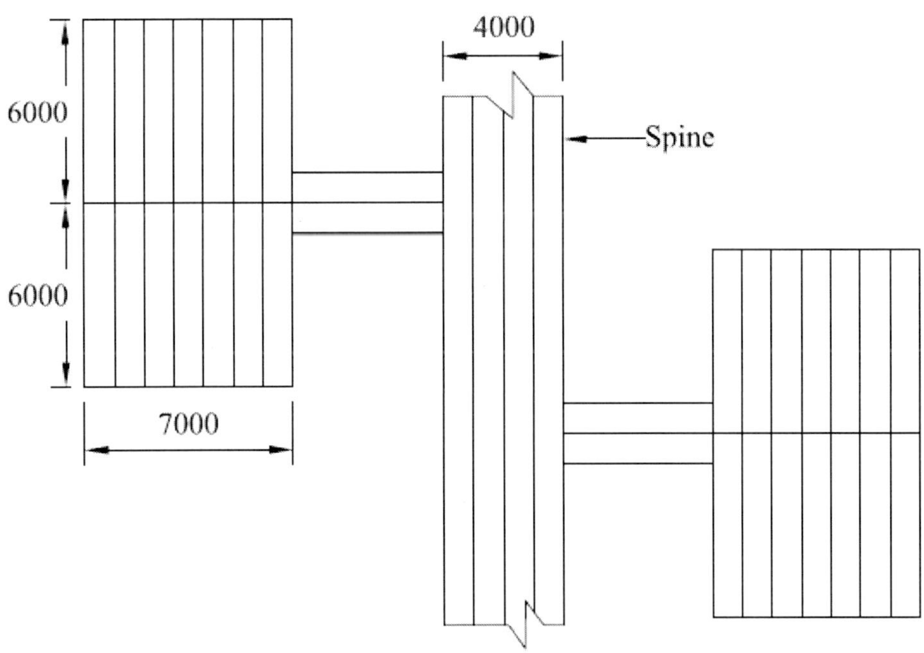

Figure 10.14. Conceptual arrangement of precast C-channels for house floor and spine construction [courtesy, author].

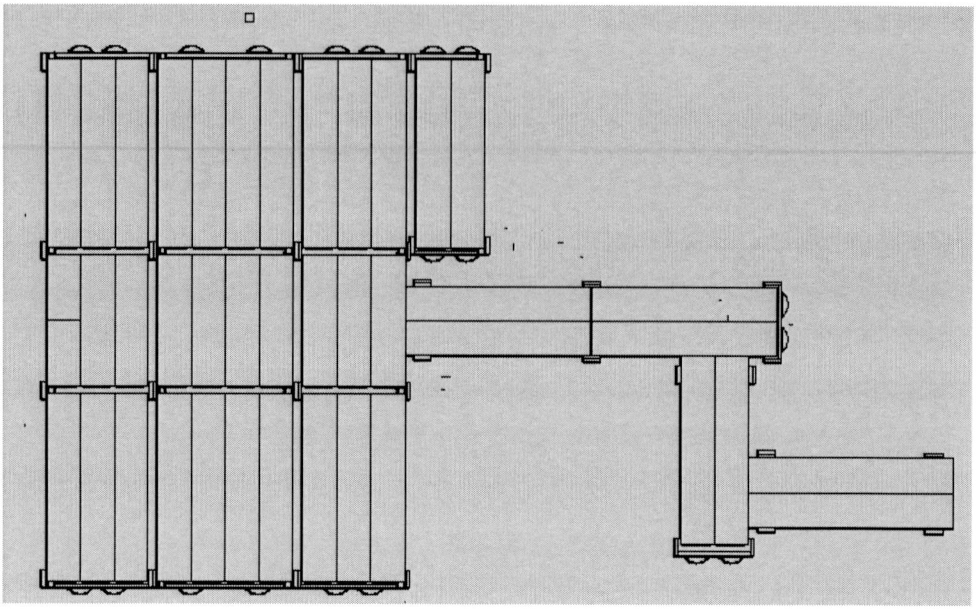

Figure 10.15. Typical plan of a resort villa and walkway using precast C-channels [5].

10.4.6. Pile Depth Determination

The brief information on pile bearing capacity, sea level, wave force, current force, lateral load resistance and scour depth is shown below for determining the pile depth in water. The design calculation can be seen elsewhere.

a) Pile Bearing Capacity

In order to determine the pile bearing capacity, it is essential for single piles that the working load be not greater than the ultimate bearing capacity divided by an appropriate factor of safety [6]. In general, an appropriate factor of safety for a single pile is between two and three. There are few methods available to determine the pile load capacity using the soil properties. Equilibrium of the pile under the applied load and resistance offered by the soil in terms of point bearing capacity and, friction and adhesion of the shaft is assumed. Thus the load is transferred to the soil partly as point bearing pressure at its base and partly as friction and adhesion along the surface of the shaft [7]. The required soil properties are:

1) Type of soil
2) Angle of internal friction of soil
3) Unit weight of soil
4) Standard Penetration Test (N-value)
5) Bearing capacity of soil
6) Average cohesion of soil

b) Sea Level

The high tide level and low tide level needs to be determined to avoid the house, spine and villa being flooded by sea water. Besides that, there must be a minimum clearance between the soffit of the floor slab and the high tide level to avoid any flood happening. High tide level is the maximum sea level during the rising tide and low tide level is the lowest sea level of the sea water flow.

c) Wave and Current Forces

The simple approach to wave force calculation on the fixed structure is to assume that the maximum wave force can be expressed as the equivalent static force caused by a solitary wave of the shape [8]. Tomlinson (2015) [8] has mentioned that the drag and inertial forces are exerted on the structure by water particles which move in an elliptical path. It is to determine the water particle velocity at any point having co-ordinates horizontally from the wave crest and vertically above the sea-bed.

The structure is affected by the velocity and direction of current. This can be done by on-site measurements using a current meter and float tracking which should include the determination of the variations in current velocity between the water surface and the sea-bed [8].

The maximum horizontal force on the pile is due to the combined wave and current forces at high tide level. At this stage of the tide, the storm wave crest is assumed at +7.5m and the storm wave trough is assumed at +5.5m.

d) Resistance to Lateral Load

The first step is to determine whether the pile will behave as a short rigid unit or as an infinitely long flexible member. This is done by calculating the stiffness factors for the particular combination of pile and soil. The stiffness factors are governed by the stiffness (EI value) of the pile and the compressibility of the soil. For most normally-consolidated clays and for granular soils, the soil modulus is assumed to increase linearly with depth.

e) Scour Depth

The scour condition is assumed to happen as live bed scour where $T_0 > T_{cr}$ everywhere on the bed and sediment is being transported by the flow from upstream, through and out of the scour hole. T_0 indicates the frictional force exerted by the flow on the bed per unit area of bed and T_{cr} is the corresponding threshold value for sediment motion.

10.5. Chapter Remarks

This chapter discusses how simple civil engineering construction can be performed in marine conditions for houses, spines etc. using IBS structural products. The precast products such as C-channels, beams, round single pile caps and round piles are the best option in marine construction for maintaining the marine ecosystem, particularly the prestine beauty.

REFERENCES

[1] *Wikipedia*. Kampong Ayer. Retrieved 2 February 2016.

[2] M. Florea. 2011. Life Red Sea project. Ecotourism in the state of Queensland-Australia. *Journal of EcoAgriTourism Biodiversity*. 7(1).

[3] *http://www.theborneopost.com/2016/02/13/solar-powered-water-purifier-promises-safe-and-clean-water/#ixzz4070Scgvs*, Retrieved 14 February 2016.

[4] *BIOGILL*. Decentralised Sewage, Retrieved 9 February 2016, Dr Tony Taylor, Resort Wastewater Treatment System using BioGill Technology. Retrieved 9 February 2016.

[5] Look Kuan Chung, 2012. Use of precast structure for marine environment, *FYP*, Universiti Malaysia Sabah.

[6] BS8004 1986: *Foundation, Section 7: Pile foundation*. London, British Standard Institution.

[7] N. S. V. Kameswara Rao, *Foundation Design: Theory and Practic*e, Publisher: Wiley, ISBN: 978-0-470-82534-1.

[8] Michael Tomlinson and John Woodward. 2015. *Pile Design and Construction Practice. Sixth edition*. CRC Press.

INDEX

#

A

B

C

D

E

F

G

H

I

J

K

L

M

N

O

P

S

T

U

V

W

Y